ALEXANDER WOODCOCK is one of the foremost experts on catastrophe theory in the United States. He has studied with René Thom and Christopher Zeeman, the theory's two most famous exponents, and frequently lectures on catastrophe theory both in the United States and abroad. Alexander Woodcock teaches biology at Williams College in Williamstown, Massachusetts.

MONTE DAVIS is a science writer who has taught biology and mathematics. He has been following the development of catastrophe theory for over eight years.

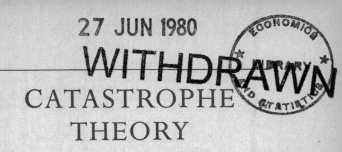
CATASTROPHE THEORY

ALEXANDER WOODCOCK
AND
MONTE DAVIS

PENGUIN BOOKS

Penguin Books Ltd, Harmondsworth, Middlesex, England
Penguin Books, 625 Madison Avenue, New York, New York 10022, U.S.A.
Penguin Books Australia Ltd, Ringwood, Victoria, Australia
Penguin Books Canada Ltd, 2801 John Street, Markham, Ontario, Canada L3R 1B4
Penguin Books (N.Z.) Ltd, 182–190 Wairau Road, Auckland 10, New Zealand

—

First published in the U.S.A. by E. P. Dutton
and in Canada by Clarke, Irwin & Co. Ltd 1978
Published in Pelican Books 1980
Copyright © Alexander Woodcock and Monte Davis, 1978
All rights reserved

—

Made and printed in Great Britain by
Hazell Watson & Viney Ltd, Aylesbury, Bucks
Set in Linotype Pilgrim

for A. N. W.

CONTENTS

ACKNOWLEDGEMENTS

FOR their time, their interest, their questions, and their answers, we gratefully acknowledge the generous assistance of:

Art Appel	Donald Beaver
Andrew Crider	Jack Cowan
Selby Davis	Chip Delany
Lee Drickamer	James Eells
Bob Friedricks	William Grant, Jr
George and Marion Goethals	Martin Golubitsky
John Guckenheimer	Victor Hill
Benoit Mandelbrot	George Marcus
Tim Poston	Mead Over
Gilbert Spencer	Regina Rouse
Susan Woodcock	Gordon Winston

— and, of course, Christopher Zeeman, and our editor, Nancy Crawford. To them and to others we owe many insights; our mistakes are our own.

A.E.R.W.
M.D.

I

A NEW KIND OF
THEORY

*'The discovery of a piece of mathematics that fits the
world in a new way is a rare event'* – TED BASTIN

CATASTROPHE theory is a controversial new way of thinking
about change – change in a course of events, change in an object's shape, change in a system's behaviour, change in ideas
themselves. Its name suggests disaster, and indeed the theory
can be applied to literal catastrophes such as the collapse of a
bridge or the downfall of an empire. But it also deals with
changes as quiet as the dancing of sunlight on the bottom of a
pool of water and as subtle as the transition from waking to
sleep.

The theory is controversial because it proposes that the
mathematics underlying three hundred years of science, though
powerful and successful, have encouraged a one-sided view of
change. These mathematical principles are ideally suited to
analyse – because they were created to analyse – *smooth*, continuous, quantitative change : the smoothly curving paths of
planets around the sun, the continuously varying pressure of a
gas as it is heated and cooled, the quantitative increase of a
hormone level in the bloodstream. But there is another kind of
change, too, change that is less suited to mathematical analysis :
the *abrupt* bursting of a bubble, the discontinuous transition
from ice at its melting point to water at its freezing point, the
qualitative shift in our minds when we 'get' a pun or a play on
words. Catastrophe theory is a mathematical language created
to describe and classify this second type of change. It challenges
scientists to change the way they think about processes and
events in many fields.

The theory's creator, Professor René Thom of France's IHES

(Institute for Advanced Scientific Studies), did not set out to provoke controversy when he began to develop these ideas fifteen years ago. He was a pure mathematician, knowledgeable in optics and other branches of science but not a specialist in any of them. He had, however, thought deeply about the order of nature, and about how it is reflected in all scientific theories. He was (and is) what used to be called a 'natural philosopher'. That term is unfamiliar today; we are used to more specialized terms such as 'scientist', coined about 1840. Before that, though, before the boundaries dividing mathematics from science and the sciences from each other were so sharply drawn, the term was common. It was as a natural philosopher, for example, that Isaac Newton established the mathematical foundations of modern science three centuries ago in his masterpiece, *Philosophiae Naturalis Principia Mathematica*.

THE APPLE AND THE STARS

It all began, the story goes, with an apple – not Eve's but the one that inspired Newton as he sat in the garden at Woolsthorpe in 1666. Of course his great achievement was not as simple as that, but we cherish the symbol : it pleases us to think that an explanation of gravity, fully ripened, fell from the tree of knowledge.

In fact, Newton's triumph was not an *explanation* of anything, but a new way of *defining* and *analysing* changes in motion. He gave definitive form to the concepts of force, mass and acceleration, and he asserted with vaulting confidence that the same principle governed the falling apple and the heavenly bodies. This was the culmination of the work of Copernicus, Kepler and Galileo : a conceptual structure that made coherent their findings, and a mathematical tool – the calculus – to deal with all continuously changing quantities. 'Newton did not show the cause of the apple falling, but he showed the similitude between the apple and the stars,' wrote biologist D'Arcy Thompson sixty years ago. 'By doing so he turned old facts into new knowledge; and was well content if he could bring diverse phenomena under "two or three Principles of Motion . . . though the Causes of those Principles were not yet discovered".'

Nor have they been discovered since. We do not know the cause of gravity, nor of the other fundamental forces that have been identified since Newton's time. Yet his principles of motion and the mathematics introduced by him and his German contemporary Leibniz, set the pattern for two centuries of discovery. The new discipline of celestial mechanics was born, and the motions of the earth, moon and planets became more predictable than clockwork. Tiny irregularities in those motions were analysed with calculus and its mathematical descendants, and the results pointed to unknown planets that were discovered just as predicted. The seemingly absolute certainty of Newton's methods offered every scientist a tempting goal : it seemed only a matter of time until all of physics, and then the other sciences, would become equally precise, quantitative and predictive. J. L. Lagrange, a mathematician who did much to extend and refine the Newtonian approach, spoke for a confident age when he remarked that surely such an intellectual advance could happen only once : after all, there was only one universe, and Newton had revealed its laws.

That certainty couldn't last, of course. The twentieth century has taught us that the universe is a queerer place than we imagined, perhaps (in J. B. S. Haldane's words) queerer than we *can* imagine. Neither the unstable atom nor the constant speed of light would fit the 'classical' scheme of Newtonian physics. A gap opened between what scientists were observing and what they could explain. On the smallest scale, change is sudden and discontinuous : electrons jump from one energy level to another without passing through states between. At high speeds, Newton's principles no longer hold : the relation of force to acceleration changes, as do mass, dimensions and time itself. Classical physics had to be amended with new, more comprehensive principles, and its mathematics supplemented with ideas that had previously seemed utterly abstract. The results were quantum physics and relativity, anti-matter and curved space.

Even in more familiar realms of experience, the belief that

all natural phenomena can be explained in terms of matter and motion, fundamental forces and continuous changes is weaker than it used to be. It is true for much of physics, and for some aspects of chemistry, a science which became firmly quantitative only in the nineteenth century. It is much less true for organic chemistry, and hardly true at all for biochemistry. Physical sciences such as geology and meteorology, in which complexity cannot be idealized away, still rely less on theory than on specialized, qualitative descriptions and judgements; the weatherman doesn't doubt Newton's principles, but he also has his own empirical laws that cannot be derived from them. In biology, with its wealth of qualitative data, mathematical theory is still in its infancy. As chemist and computer theorist Christopher Longuet-Higgins put it, 'One could state with great accuracy the laws describing the motion of a falling apple, but people hadn't yet explained exactly how a pip grew into a Cox's Pippin.' And in psychology and social science, quantitative prediction is a remote dream.

Yet the successes of physics, and their technological consequences, have been so great that many people still contend that other disciplines are truly scientific only to the extent that they are modelled on the Newtonian pattern. They insist that given enough data and enough time, we can pass systematically from fundamental physics to chemistry, then to biology, then to the brain and society. Only then will the bits and pieces of qualitative thinking fall into place. 'All science is either physics or stamp collecting,' atomic experimenter Lord Rutherford told his students early this century. 'Qualitative is but poor quantitative.'

This attitude persisted even after Rutherford's own discoveries helped to usher in the abstractions and paradoxes of modern physics, with its mathematical models in which neither matter nor motion can be visualized. Quantum physicist P. A. M. Dirac wrote, 'The main object of physical science is not the provision of pictures, but the formulation of laws ... if a picture exists, so much the better; but whether a picture exists or not is a matter of only secondary importance.'

To René Thom, although he is no stranger to mathematical abstractions, Dirac's statement signals resignation rather than comprehension. Thom believes that the provision of some kind of picture, at least to the mind's eye, is of primary importance. To him, our qualitative grasp of form and geometric order goes deeper than our quantitative grasp of number and magnitude, and should not be relinquished. In response to Dirac, he wrote:

... I am certain that the human mind would not be fully satisfied with a universe in which all phenomena were governed by a mathematical process that was coherent but totally abstract. Are we not then in wonderland? In a situation where man is deprived of all possibility of ... interpreting geometrically a given process, either he will seek to create, despite everything, through suitable interpretations, an intuitive justification of the process, or he will sink into resigned incomprehension which habit will change to indifference. In the case of gravitation there is no doubt that the second attitude has prevailed, for we have not, in 1975, less reason to be astonished at the fall of the apple than had Newton.

Newton himself knew he had said nothing about the cause of gravity. 'I do not feign hypotheses,' he replied when pressed on that score. Today we know of four fundamental forces – gravity, electromagnetism, and two different forces within the atomic nucleus – but all are mysterious. A student who asks 'Why?' is usually told that force is an explanatory principle, and cannot itself be explained. He learns that if physicists are not exactly 'indifferent' to the question, they are certainly resigned. At best, they hope for a unified theory to combine the four mysteries into one. Looking closer, the student sees that visualization has always played an important role in dealing with the mysteries. Newton's work in physics was largely geometric in spirit; James Clerk Maxwell, whose mathematics transformed the study of electromagnetism, described its forces in terms of a field spreading throughout space; and even the greatest modern physicists (Einstein was no exception) frequently visualize forces as 'hills' and 'valleys' on a map of space-time. As Thom puts it, 'The dilemma posed to all scientific explanation is this: magic or geometry.' Either we must stop asking 'Why?' or we must seek to extend our intuition of form to new levels, to see that processes and events have shapes of their own.

It is possible to see the shapes of processes within the traditional framework, but only for a certain class of processes – those involving continuous change. In calculus, the relationship between two quantities can be expressed as a set of points on a graph, each standing for a certain level of x, a certain level of y. If those points form a smooth, continuous curve – the kind mathematicians call 'well-behaved' – then calculus and the analytic techniques descended from it make it possible to analyse the process, determine its rate of change at an instant and its total change over a period of time, and summarize it as an equation relating x and y. We can speak of a process 'bottoming out' or reaching its 'point of diminishing returns', and we understand at once the ominously increasing slope of the population curve.

As we have seen, though, much of reality is not so obliging. Many processes yield graphs with obstinately ill-behaved curves : there are spikes, breaks and regions where one value of x corresponds to any of several values of y, or vice versa. The planets travel in stately Newtonian paths, but meanwhile winds wrap themselves into hurricanes, chickens alternate with eggs, and we change our minds. Discontinuity is as much the rule as the exception. To take a relatively simple example, the physical properties of water are discontinuous at the freezing and boiling points. A graph of its temperature versus the flow of heat energy shows large, abrupt thresholds at those points, and no simple equation can relate the two quantities. It is not surprising, then, that there is no satisfactory physical theory of such phase transitions, although we know a great deal about water molecules and the forces between them. How does water differ in this respect from glass, which softens steadily with increasing temperature, or from carbon dioxide, which can change directly from 'dry ice' to a gas? It may be that a mathematical approach developed for continuous change is not the best way to understand such processes.

What kind of mathematics does Thom offer instead? The answer is topology, a sophisticated descendant of geometry. Instead of the straight lines, restricted curves and regular solids of Greek geometry, topology deals with all conceivable forms, abstract and multi-dimensional forms as well as those that can be drawn. Thom is an acknowledged master of differential topology, a specialized field which combines those forms with elements from calculus in order to deal with questions of stability and transformation. Greek geometry was essentially timeless: any triangle or circle in the real world was considered an imperfect, changeable 'shadow' of ideal, eternal, mathematical form. Thom uses differential topology to start from the opposite premise: that changes of form (in processes as well as objects) are real, and that the aim of science is to grasp what he calls the universe's 'ceaseless creation, evolution and destruction of forms'.

Because of its foundation in topology, catastrophe theory is qualitative, not quantitative. Just as geometry treated the properties of a triangle without regard to its size, so topology deals with properties that have no magnitude; for example, the property of a given point being inside or outside a closed curve or surface. This property is what topologists call 'invariant' – it does not change even when the curve is distorted. A topologist may work with seven-dimensional space, but he does not and cannot measure (in the ordinary sense) along any of those dimensions. The ability to classify and manipulate all types of form is achieved only by giving up concepts such as size, distance and rate. So while catastrophe theory is well suited to describe and even to predict the shape of processes, its descriptions and predictions are not quantitative like those of theories built upon calculus. Instead, they are rather like maps without a scale: they tell us that there are mountains to the left, a river to the right and a cliff somewhere ahead, but not how far away each is, or how large.

These are serious limitations for the theory, but then, Thom does not suggest that his ideas should replace the quantitative approach, only that they can do justice to aspects of the world which it tends to short-change. The maps produced by existing theories, after all, make the world a patched-together place of smooth curves and warning signs: 'Beware Threshold', 'Ill-Behaved Process – Theorize at Your Own Risk' and so forth. Catastrophe theory is a way to see what lies beyond.

THE STABILITY OF CHANGE

A second theme of catastrophe theory, which again reveals Thom as a natural philosopher, is its adaptation to long-standing questions about the forms that recur time and again in nature. Many people have wondered, for example, at the similarity of branching patterns in a tree, a river system and a nerve cell. How does this qualitative similarity emerge from three very different sets of circumstances? Is it coincidence, or an indication of a common principle at work in each case?

Lord Rutherford, were he still alive, would probably dismiss the question as more stamp collecting. Let's find out all we can about botany and fluid dynamics and neurophysiology, he would say, and break it all down to moving particles and changing forces, and then we'll see. But Thom takes a different view:

The choice of what is considered scientifically interesting is certainly to a large extent arbitrary. Physics today uses enormous machines to investigate situations that exist for less than 10^{-23} second, and we surely are entitled to employ all possible techniques ... But we can at least ask one question: many phenomena of common interest, in themselves trivial (often to the point that they escape attention altogether!) – for example, the cracks in an old wall, the shape of a cloud, the path of a falling leaf or the froth on a pint of beer – are very difficult to formalize, but is it not possible that a mathematical theory launched for such homely phenomena might, in the end, be more profitable for science?

How far would a traditional analysis take us in explaining the cracks in an old wall? They appear in response to many factors: the varying strength of the brick and mortar in different places, changes in humidity, even the soil settling beneath the wall. Together, these factors produce stresses that interact over the years with such complexity that it would be impossible, even taking the most precise measurements and using the largest computer, to point to a new wall and say, 'Cracks of such-and-such a size will appear precisely there, there and there on the following dates . . .'

Yet wherever the cracks appear, they show a tendency to extend towards each other, to form characteristic networks, to form specific types of junctions. The location, the magnitude and the timing of the cracks (their quantitative aspects) are beyond calculation, but their patterns of growth and the topology of their joining (the qualitative aspects) recur again and again.

And the path of a falling leaf? It depends on every detail of the leaf's curvature and outline, for they determine the air resistance that the leaf encounters. It can be altered by the slightest breeze, even by minute fluctuations in temperature and humidity on the way from the twig to the forest floor. Only a foolhardy scientist would attempt to predict where a given leaf will land, let alone its path. Yet anyone who has walked in a New England forest in autumn knows the slow, side-slipping descent, like a series of shallow 'U's, of maple leaves, and the fluttering spiral of birch leaves. No two leaves, no two paths, are quantitatively the same, but we can distinguish and recognize them by their qualitative behaviour.

It might be argued that what we distinguish or recognize is a matter for psychology, not for mathematics. But for Thom, it is a sign of a special kind of stability in the processes themselves. 'Almost any natural process', he argues, 'exhibits some kind of local regularity . . . which allows one to distinguish recurrent identifiable elements denominated by words. Otherwise, the process would be entirely chaotic and there would be nothing to talk about.'

These 'recurrent identifiable elements' can be characteristic shapes, like that of a snowflake or a butterfly. Or they can be characteristic stages of a dynamic process, like the formation of snowflakes from water vapour or the metamorphosis that turns a caterpillar to a butterfly. In either case, they have the property Thom calls 'structural stability'. Their qualitative features are recurrent, even though the circumstances giving rise to those features are never exactly the same in quantitative terms. As Thom points out, all science is based on an implicit assumption of structural stability. Two experiments will never yield exactly the same quantitative results, because experimental conditions cannot be duplicated precisely, and external disturbances cannot be completely eliminated. It is not surprising if a second experiment produces measurements a little greater or less, graphic curves a little steeper or shallower, than the first. But if the curve bends down instead of up, if the liquid that froze in the first experiment boils in the second, something is seriously wrong. Science is possible only if observations and results are *qualitatively* repeatable.

Thom's goal is to describe the origin of forms, which he calls *morphogenesis*, borrowing a word used by the Greeks and also by modern biologists. To do so, he has created a mathematical language – catastrophe theory – built on the assumption of structural stability, and stressing qualitative rather than quantitative regularity. He believes that it is general enough to fit the snowflake, the butterfly, and the processes that shape them, as well as the more complex and highly organized processes by which the words 'snowflake' and 'butterfly' enter our minds and speech.

Existing theories go only a little way in explaining morphogenesis. For a simple process, one quantitative approach relies upon statistics, treating large-scale form as the average or most likely outcome of many random events. The heap of sand in the bottom of an hourglass is an example of such a statistically determined form. But the more complex and highly organized the process, the less satisfactory this explanation becomes. Can averages and probabilities account for the exquisitely precise unfolding of a fertile seed – with soil, water, air and light – as it develops into an apple tree? And if they can, how can they account for the tree's creation of new seeds? It is easy to see how a crystal dissolves as conditions change, for example, but the apple seed's development is not reversible : it is a stable one-way series of changes.

Another quantitative approach relies upon nature's tendency to maximize or minimize certain quantities, as a soap bubble minimizes surface tension by assuming a spherical shape with maximum volume. These principles are undoubtedly operative in the development of the apple tree, too, but there they are so complex as to defy calculation. The tree appears to offer a maximum of leaf area to the sun, but to minimize water loss to the air; to use a minimum of chemical energy to carry out its metabolic functions, but to maximize its chances of reproduction, even if all but a few seeds are wasted.

In response to such complexity in nature, the traditional tendency of quantitative science has been to seek a detailed control mechanism. Many scientists believe that we have found it in the genes, which direct the synthesis of every organism's structural and metabolic proteins. But here, too, we encounter the gap between principle and practice that has kept complex sciences from becoming as precise as celestial mechanics. In the words of John Tyler Bonner, an expert on biological morphogenesis : '... there have been great advances in how one gets from a particular gene to a particular protein it is coded to design, but now we are asking how this protein, or how many such proteins, produce a consistent gastrulation [the reshaping of a ball of multiplying cells into an embryo], or a perfectly proportioned leg, or an eye. We even need to have it direct the whole structure and composition of an ant society. The gap between those first gene products and such complicated end results is the gap that lies yawning before us; this is the sensitive spot.'

In Thom's view, the most striking feature of such processes is not their quantitative complexity, but their qualitative stability. Somehow a great many processes are gathered into a clear-cut result, even under considerable quantitative variation. An apple seed may experience a wide range of temperature, moisture, soil acidity and so on, but if it grows at all it will grow into an apple tree, not a cactus or a cattail. Even a gene mutation can often be suppressed by the inherent stability of biological processes, just as the apple tree can maintain its function in many possible settings of soil and climate.

Biologists and physiologists have long recognized the vital ability to preserve what they call *homeostasis* (Gr. 'same state'). For example, the kidneys are adept at keeping the fluid and salt levels of the blood within a narrow range. One of the foremost biologists of this century, C. H. Waddington, coined the word *homeorhesis* (Gr. 'same path') for processes of biological development which pursue a stable course of change. Waddington subsequently played an important part in the growth of catastrophe theory, and his influence is clear when Thom asserts that life is a process in which stability, not merely a given arrangement of genes, is passed on : 'This set of regulating mechanisms [homeostasis and homeorhesis], always the same for each individual of the species, is what we should consider the genetic endowment.'

From this perspective, the chemical stability of the genes and the ecological stability of a species are part of the same theory. Thom extends this notion of stability to the forms of inorganic systems as well. 'If sodium and potassium exist,' he reasons, 'it is because there is a corresponding mathematical structure guaranteeing the stability of their atoms; such a structure can be specified, in quantum mechanics, for a simple object like the hydrogen molecule, and although the case of the sodium or potassium atom is less well understood, there is no reason to doubt its existence. I think that likewise there are formal structures ... in biology which prescribe the only possible forms capable of having a self-reproducing dynamic in a given environment.'

The most complex system imaginable is the mind – by definition, since the mind must be at least one degree more complex than whatever it imagines. Catastrophe theory proposes that qualitative stability is a necessary attribute of thought; without it, recognition and memory would be impossible. Thom holds that our concepts are mathematical models, topological maps, of the objects and processes that inspire them. While some mathematicians insist that their ideas are totally free creations, Thom takes an evolutionary view : just as our bodies are adapted for crawling, walking and running, and just as our hands are adapted for grasping objects and shaping tools, so our minds are adapted for modelling topologically the world in which bodies, hands and minds evolved.

Discontinuity and qualitative change occur everywhere in thought, language and perception. When you look steadily at this optical illusion, the Necker cube ...

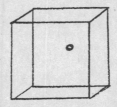

... the transition from one way of seeing it to another is discontinuous. You cannot stop it halfway; you cannot catch the little circle in motion from the centre of one face to the corner of another. There is no way to predict how the figure will appear at first glance, but the reversal after a few seconds is always the same. Whatever the perceptual mechanisms at work, the change itself is stable. Both visual interpretations are coherent. Both make sense of the pattern.

The eventual position of catastrophe theory itself may be comparable. It offers an alternative way of looking at the world – not more correct than Newton's way, perhaps more complete, surely radically different. It points out qualitative similarities in a wide variety of processes, just as the analogies of ordinary language do, but with the advantage that its analogies can be rigorously classified and combined, using mathematics as well adapted for the purpose as Newton's calculus was for analysing quantitative relationships. For three hundred years we have explored the world using maps of those relationships. Now, with new maps, there is a chance to see new territory : the landscapes of change.

THE UNFOLDING OF
CATASTROPHE THEORY

'The imaginative ideas from which scientific work originates depend on a sensitive receptiveness to the oddity of nature, essentially similar to that of the artist. When they are first proposed they often have the same quality of unexpectedness, and perhaps wrongheadedness, as say, cubism, abstract art, or atonal music' – C. H. WADDINGTON

A NEW idea doesn't necessarily meet with a warmer reception in science than in art or any other human activity. Like most of us, scientists are generally content with established ideas unless a specific problem demands new mathematical or conceptual tools. They rarely have contact with pure mathematicians, and most of them are no more concerned with the philosophy of science than most lawyers are with the philosophy of law.

Although René Thom's creativity in pure mathematics earned him honour among his colleagues twenty years ago, it was familiar only to a small group. And although his natural philosophy was profound and original, it would not by itself have made a stir in science. Yet combined in catastrophe theory, the mathematics and philosophy are having a wide impact. Thom's book, *Structural Stability and Morphogenesis*, was published only a few years ago, but today catastrophe theory is being used to describe phenomena as diverse as psychological crises and chemical reactions. At the same time, critics are denying that the theory can or should be applied at all. It's beautiful mathematics, they say, but has nothing to do with the real world, and the claims that have been made for its wide range and possible predictive value are 'the height of scientific irresponsibility'. In an exchange of letters in *Science*, mathematician Marjorie Senechal wrote: 'I am fascinated by the sense of mission that drives some scientists to denounce heretics and heresies, and by the ease with which others abandon their objectivity and jump on bandwagons.'

How did Thom's abstract new vision of the world so quickly became a heresy, and the opposition to it a bandwagon? And how was his work in an esoteric branch of mathematics turned to detailed applications? To answer these questions, we must look back at the theory's roots in mathematics and science; at its growth from the early 1960s onwards, which brought it to the attention of many scientists even before the publication of Thom's book; and finally, at its divergent evolution in Thom's thinking and that of E. Christopher Zeeman, the English mathematician who contributed to catastrophe theory's birth, has been its most active advocate, and stands at the centre of the current controversy.

THE ROOTS OF THE THEORY

Thom has long had a reputation for the ambition, even the riskiness, of his mathematical ideas – and for their success. In 1946, at twenty-three, he graduated from the prestigious École Normale Supérieure. This college, long a pinnacle of France's national school system, awards only a few dozen degrees each year in science and mathematics. In 1951 Thom wrote his doctorate in topology. 'He was full of ideas and enthusiasm,' recalls a contemporary. 'There were great topologists in the older generation, but Thom wasn't anyone's disciple – he had more in common with Darboux and Poincaré.' (Gaston Darboux was France's leading geometer at the end of the nineteenth century – a century in which Gauss, Bolyai and Lobachevsky had created alternatives to classical, three-dimensional, Euclidean geometry, and Riemann had generalized their work into a theory that paved the way for Einstein. Henri Poincaré, Darboux's contemporary, has been called 'the last universalist': the last mathematician to do first-rate work in all areas of pure and applied mathematics, and to write for a lay audience as well. Many believe that he very nearly anticipated Einstein in stating the theory of relativity.)

Thom published few papers, although it is said that they contained many of the most important ideas in topology of the 1950s. His colleagues prodded him to get his work on paper, and he gained a reputation for preferring intuition to technical rigour. 'Some mathematicians go at their work like engineers building a six-lane highway through the jungle,' says mathematician Tim Poston, 'laying out surveying lines, clearing the underbrush and so on. But Thom is like some creature of the mathematical jungle, blazing a trail and leaving just a few marks on his way to the next beautiful clearing.'

In a 1954 paper, Thom set forth the concept of transversality. To squeeze it into a nutshell, transversality concerns the ways in which the smooth curves of analysis (the abstract descendant of calculus) can intersect or 'cut' each other. These cuts can be clean, or they can be mathematically messy in a variety of interesting ways; transversality demonstrated that most cuts are clean, and gave mathematicians a 'handle' on those that aren't. Many problems considered as part of analysis were illuminated by Thom's topological approach, and transversality has stimulated and unified a number of mathematical advances since 1954. It made possible Thom's theory of co-bordism, for example, for which he won the 1958 Fields Medal (mathematics' highest international honour), and other developments such as Milnor's 'exotic sphere' – a seven-dimensional form with properties as surprising to a topologist as, say, those of a fire-breathing dragon would be to a biologist.

Thom taught at the University of Grenoble until 1957, then moved to the University of Strasbourg. He pursued the implications of transversality and linked them to the work of others such as Hassler Whitney, an American topologist at Princeton's Institute for Advanced Studies. Whitney had studied the singularities of mappings – phenomena that occur when the points of one surface are projected on to another as the surfaces are topologically distorted.

Besides its meaning in topology, 'singularity' (or stationary points) has another meaning in calculus and analysis. There, it is a point on a graphic curve where the direction or quality of curvature changes. In Figure 1 the four singularities (a local maximum, two local minima, and a point of horizontal inflec-

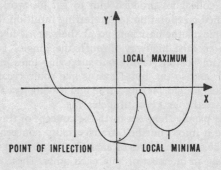

Figure 1. Singularities of a curve

tion) are labelled: you can see that at each one, the slope of the curve is momentarily level. The maximum and the two minima are called 'local' because they are not necessarily the highest or lowest points on the curve, merely higher or lower than their immediate neighbours. These points are of interest in many practical applications of calculus – for example, if x represents the temperature at which a fuel is burned and y the amount of pollution; or if x represents the amount of pressure used to forge metal and y its resulting strength. There is a discipline known as the calculus of variations, in which mathematicians have developed general techniques for locating these singularities (given the equation corresponding to the curve). Thom became interested in the relationship between the calculus of variations and Whitney's topological singularities.

He was not the first to see such a connection. In the 1880s and 1890s Henri Poincaré had linked calculus and topology (then called 'analysis situs', analysis of location) to create *qualitative* dynamics and apply it to unsolved problems of planetary motion. This may seem strange; after all, dynamics had been a firmly quantitative field since Newton. But Newton's methods yield explicit solutions only for the interaction of two bodies – for example, the sun and the earth, or the earth and the moon. When three or more bodies are involved, the equations of motion cannot be solved directly, and even approximate solutions require tedious, complex procedures. Around 1800 Pierre Simon de Laplace, the great mathematical physicist known as 'the Newton of France', had tried at length – but without success – to show that all the two-body attractions of the solar system added up to a stable dynamic system, a grand perpetual-motion machine that would run forever.

Poincaré set out to show that even if quantitative solutions were impossible, it was still possible to make progress on important questions : does a complex, many-body system return periodically to the same arrangement? Does a slight perturbation simply 'nudge' the whole system, or does it lead eventually to qualitatively different behaviour, such as a planet spiralling into the sun or colliding with another planet? Though he did not fulfil Laplace's earlier hopes, Poincaré inaugurated a valuable new approach. His fellow mathematicians saw its value, but thought it arbitrary because it was adapted to a particular physical problem rather than being part of a general method : they called it 'so new, so perfect ... and so hard to follow up'. Gaston Darboux, Poincaré's biographer, explained : 'Poincaré was an intuitionist. Having once arrived at the summit he never retraced his steps. He was satisfied to have crashed through the difficulties, and left to others the pains of mapping the royal roads destined to lead more easily to the end.' As E. T. Bell, a historian of mathematics, summed it up in 1937 :

Much of Poincaré's work in his astronomical researches was qualitative rather than quantitative, as befitted an intuitionist, and this characteristic led him, as it had led Riemann, to the study of analysis situs. On this he published six famous memoirs which revolutionized the subject as it existed in his day ... He *modernized* the attack [on planetary motions]; indeed his campaign was so extremely modern to the majority of experts in celestial mechanics that even today, forty years or more after Poincaré opened his offensive, few have mastered his weapons and some, unable to bend his bow, insinuate that it is worthless in a practical attack.

That it was not worthless became clear in that same year, when two Russian mathematicians, Andronov and Pontryagin, carried on Poincaré's ideas in their general definition of structural stability. They made mathematical questions out of Poincaré's physical ones. Given the equations describing any dynamic system, they said, the crucial question was how the stable solutions for those equations were distributed topologically. Was a stable state of the system part of a continuous range, or an 'island' surrounded by instability? Would a small quantitative change alter the solutions slightly, or produce very different new ones, or perhaps leave none at all? At about the same time, American topologist Marston Morse was renewing the topological approach to the calculus of variations, making it possible to find the maxima or minima of whole families of curves. His work can be traced back to Poincaré and forward to Hassler Whitney, his younger colleague and successor at the IAS.

So Thom's combination of topology and analysis was not unprecedented, nor was his interest in physically stable and unstable processes. He turned over in his mind the meaning of structural stability, examining singularities of higher dimension than those Whitney had considered. He experimented with the optical phenomena called 'caustics', patterns created when light is reflected by imperfect mirrors or refracted by imperfect lenses such as airborne droplets of water (the shape of the rainbow is determined by caustics). At the edge of a caustic, the light intensity reaches a maximum value and suddenly drops off. To Thom, caustics were striking visual unfoldings of singularities in the equations of optics – singularities that reappeared in shock waves and other discontinuous phenomena, generating similar shapes again and again.

Topology and stability were also on the mind of E. Christopher Zeeman, then a lecturer in mathematics at Cambridge University. About 1960 Zeeman published several papers suggesting that topological models could bridge the gap between the quantitative findings of neurophysiology and the qualitative descriptions used in psychology. Could this kind of brain modelling, Zeeman asked, describe the fusion of impulses from thousands of distinct light-sensitive cells in the retina into a continuous, coherent visual field?

Slightly younger than Thom, Zeeman had served as a flight officer in the RAF before entering Cambridge as an undergraduate. He received his doctorate there in 1954, specializing in a branch of topology called 'knot theory', a subject even more tangled than it sounds, especially when the question is how to 'untie' a ten-dimensional knot by manipulating it in sixteen-dimensional space. Zeeman's early speculation about brain modelling was half-serious; he would write in 1976 that the papers he wrote on it 'were not very good but had the virtue of turning René Thom's attention to biology and being partly responsible for the creation of catastrophe theory'. In Zeeman's ideas, and to some extent those of contemporary thinkers such as structuralist Claude Lévi-Strauss and Noam Chomsky, who were transforming anthropology and linguistics, Thom found corroboration for his growing belief that thought and language are shaped by deep principles of structural stability just as surely as physical processes.

Beginning at mathematical conferences, a friendship grew between Thom and Zeeman. Though they were disparate physically and temperamentally – Thom stocky, crew-cut, outwardly formal and reserved; Zeeman taller, with bushy beard and hair, enthusiastic and voluble – both were intellectual mavericks. They shared, for example, a belief in the importance of spatial intuition that led Thom to oppose the replacement of geometry by set theory and algebra in the 'new maths' for lower schools. They also shared a fascination with the variety and recurrence of form in nature.

In 1963 Thom left Strasbourg to join the IHES at Bures-sur-Yvette near Paris. The Institute has no undergraduates, but the high calibre of its members draws graduate students and visiting faculty from all over the world. Its wooded paths, where small strolling seminars take place in good weather, reminded one visitor of scenes from a Hermann Hesse novel; 'it's a little surreal', he recalls, 'to look at the note board and see "automorphic groups, 3 p.m., Bois Ste-Marie"'.

The following year Zeeman went from Cambridge to Coventry, where he had been engaged to head the mathematics department of the new University of Warwick, scheduled to open in 1965. But by the time Warwick opened, Zeeman had founded a Mathematics Institute which was already well-established and growing, with a nucleus of faculty lured from Cambridge. At first the Institute consisted chiefly of a sign in front of Zeeman's office, a converted private home. Since then it has become a research centre of international stature, and its annual conferences have drawn many of the world's finest mathematicians. Although bustling, industrial Coventry may not be able to match the charms of Paris or of Rio de Janeiro, home of the equally prestigious Instituto de Matematica Pura e Aplicada, the University is set in rolling farmland a few miles outside the city, with the Mathematics Institute slightly apart on a hill. 'Zeeman has been a genius at administration and at getting support for the Institute,' says an admiring colleague. 'The 1960s were a time of growth for all the new universities in England, but Warwick stands out, and one large factor in that has been the enormous success of the Mathematics Institute.'

THE GROWTH OF THE THEORY

Another English scholar played an important part in the unfolding of catastrophe theory. He was C. H. Waddington, professor of animal genetics at the University of Edinburgh and president of the International Union of Biological Sciences. Waddington had anticipated important parts of Thom's biological thought, and was the first scientist of great stature to acclaim catastrophe theory.

In a long and distinguished career which spanned nearly half a century, from 1930 until his death in 1975, Waddington had worked in paleontology, embryology, genetics and evolution. His first research was a study of ancient ammonites (spiral-shelled sea animals related to the chambered nautilus), in whose fossils, he wrote, 'the whole developmental process is preserved so that one cannot avoid examining it'. His subsequent career can be seen as an exploration of development in the broadest sense – the development through time of species, of embryos before birth and of genetic traits.

In the 1930s Waddington was an experimental embryologist at Cambridge. He was the first researcher to maintain mammalian embryos in the laboratory. With Joseph Needham, Jean Brachet and others, he studied the processes of morphogenesis that transform an apparently uniform ball of cells into a layered structure of differentiated tissues. These investigations revealed much about the chemical signals of morphogenesis, including the surprising discovery that many substances – even some not normally found in organisms – can act as triggers for the same complex sequence of events. But Waddington kept in mind that biochemistry alone cannot explain morphogenesis. Years later he wrote: 'In the development of an embryo something determines that a group of cells will form some part of a complex organ, such as a leg rather than an arm, before it is decided whether those cells shall be bone or muscle. Now I think that this is a very astonishing performance. What kind of material substance can there possibly be which characterizes the whole leg with all its bones, muscles, nerves, blood vessels and so on, and distinguishes it from an arm with its bones, muscles and so on?'

Waddington's research convinced him that many biological processes had the property he called homeorhesis: that is, they were stable, 'canalized' pathways of change that resisted disturbing influences, like streams confined by their banks. He came to think of development as a landscape, with these pathways of change separated by higher or lower ridges that could be shifted by external conditions. He carried the same image into his later studies of genetics and evolution, showing that a greater or lesser stability of development and metabolism was an inherited trait; that natural selection worked to alter the contours of the many-dimensional landscape. Although he was internationally famous as an expert on morphogenesis, Waddington always felt the inadequacy of existing language in that field. In 1962 he wrote : 'I suggested some years ago (1940) the desirability of a theory of a generally topological kind, which would be appropriate to biological forms. I suggested that such a theory would have to be in terms of "topological operators", that is notions such as folding on a line, piercing of holes ... However, no such theory has yet been developed, and we shall have to do our best to distinguish form and shape without its aid.'

As he wrote, René Thom was well on the way to just such a theory. A museum display of embryological models caught Thom's eye, and looking at them he recognized again the shapes of the unfolding singularities in his mathematics. Each of the solid, three-dimensional models was a 'cross-section in time' of an integrated process, just as an optical caustic on a screen was a cross-section of a mathematically determined sheaf of light rays. Thom read widely in embryology, and found that Waddington and others had prepared the way for a topological approach. To him, the mathematics of stable pathways of change and the mathematics of biological form were the same, for every organism's form represents a partial record of the processes of development and metabolism. Where Waddington had visualized a qualitative change – the initial differentiation of an arm or leg, or the appearance of a new inherited feature – as a stream flowing into a new channel, Thom saw it as the emergence of a new singularity which shaped the subsequent course of the process.

For years, Thom found, ideas that were topological (although not always recognized as such) had been common in all areas of biology. As early as 1917 D'Arcy Thompson had shown that the shape of a fish or of an animal's skull, drawn on a rectilinear grid, could be altered by a continuous, smooth transformation to that of a related fish or of a skull in the animal's evolutionary predecessor. It proved impossible to develop quantitative mathematics for this remarkable visual relationship, but Thompson's *On Growth and Form* – a wide-ranging and beautifully written exploration of the mathematical and physical aspects of natural form – has had a pervasive influence on three generations of scientists. Thom refers frequently to Thompson in his own book.

In 1941 the German physicist Bernhard Bavink urged biologists 'to place the concept of measurable and countable quantity in second place, and the basic biological concept of form or *gestalt* in first place'. Bavink envisaged a mathematics of form developing from the calculus of variations and from group theory, and in fact the transformations of Thom's topology are related to group theory as well. In the 1950s American developmental biologist Paul Weiss observed the regularly spaced folds in mitochondria, the energy-producing structures within all animal cells, and proposed that their arrangement was governed by the maximum or minimum values of a periodic chemical reaction. In these 'crests and valleys of conditions favouring [the assembly of the folds]', he concluded recently, 'we encounter the phenomenon of the emergence of singularities in a dynamic system – unique points, or lines, or planes . . .'

So catastrophe theory, as it emerged in Thom's mind, was above all intended as a mathematical language for biology, although not for biology alone. In fact, Thom thought, it might be possible to make more progress in biology than in many problems of physics where sudden changes of form were hard to explain, such as the feathery growth of frost crystals or the breaking of waves: 'it may happen that biological morphogenesis, which is better known, which takes place slowly, and which is controlled strictly, may help us to understand the more rapid and fleeting phenomena of inert morphogenesis'.

All the foundations of catastrophe theory were in place by 1964. One crucial step remained to be taken, though: the establishment of a basic vocabulary for the new mathematical language. Thom had worked out thoroughly the relationship of topological singularities to the maxima and minima of calculus. He could see how the former would 'unfold' into arrangements of the latter, imposing a structure on them. And to know the structure – the arrangement – of the maxima and minima of a dynamic process would be to know its qualitative behaviour. But how many topologically different structures were possible?

Because of the persistent recurrence of similar forms which Thom had observed in nature, he believed that, at least for simple processes, there was also a limited number of archetypal structures. He was certain that there must be a unique unfolding for each singularity in such cases, but he could not prove it. He persuaded another French mathematician, Bernard Malgrange, to attack the problem. Malgrange was dubious at first, but Thom's insistence carried the day: Malgrange proved the uniqueness of unfoldings in 1964. Armed with that and his own transversality theorem, Thom reached a remarkable conclusion in 1965: that for a very wide range of processes, only seven stable unfoldings, the seven 'elementary catastrophes', are possible.

The unfoldings are called catastrophes because each of them has regions where a dynamic system can jump suddenly from one state to another, although the factors controlling the process change continuously. Each of the seven catastrophes represents a pattern of behaviour determined only by the *number* of control factors, not by their nature or by the interior mechanisms that connect them to the system's behaviour. Therefore the elementary catastrophes can be models for a wide variety of processes, even those in which we know little about the quantitative laws involved. This is an extraordinary idea: how is it possible that two processes can have features in common even when they are on different physical scales, operate under different quantitative laws and are affected by different sets of causes?

The elementary catastrophes will be presented in detail in the next chapter, but to answer this question we might compare them to the most basic regular forms of geometry. The Greeks discovered that of all possible regular polygons (two-dimensional shapes with equal sides), only three (the triangle, the square and hexagon) can be packed edge-to-edge to fill the plane. That is a mathematical restriction on anyone who tiles a wall or floor, for example, and it has nothing to do with the material of the tiles or how they are applied. The Greeks also found that if the regular polygons are assembled as the faces of three-dimensional solids, only five such solids can be constructed. And these polygons and solids appear throughout nature, in snowflakes and diatoms and crystals and honeycombs – not because geometry dictates to nature, but because there is no other way for certain natural processes to turn out. In the eighteenth century scientists were astonished to discover that the cells of a honeycomb came very near to using an absolute minimum of wax to enclose a given volume; did the bees' instincts permit them to solve a problem in the calculus of variations? Today we realize that the pressure of the bees' bodies as they work the warm, soft wax is enough to account for the phenomenon. Each cell wall is under pressure from both sides, and it assumes a shape that equalizes that pressure as much as possible. The cells are hexagonal in section for the same reason that a tray of pennies, if it is shaken until the pennies are close-packed, shows a hexagonal array. It does not matter that the movements of the bees are very different from, and far more complex than, the vibration of the tray. The qualitative, geometric pattern that results is the same. Thom believed that similarly the qualitative, topological patterns of behaviour seen in the elementary catastrophes must recur in many processes.

Mathematically, this meant demonstrating that these seven abstract forms existed, that they were unique and that they were structurally stable. Thom satisfied himself on this score and plunged ahead, although he did not give a rigorous proof (it was carried out by John Mather and by other singularity theorists in 1967–8). By 1966 Thom had finished an early draft of his book. There were prolonged delays in its publication, due chiefly to the publisher's business difficulties, which led to its acquisition by another firm. Another factor may have been the difficulty of classifying the book: exactly what was it? It was not a technical work in topology, with its philosophical asides and its speculation in biology, linguistics and psychology. Nor was it a textbook, for its subject was not part of any curriculum. Nor was it a book for the general public, because its forbidding passages about 'operators induced by a diffeomorphism $h: X \to X$ of base spaces' and the like were guaranteed to deter all but the most tenacious readers.

During the six-year delay before publication, Thom continued to update the book. He had already found a more direct way to reach a scientific audience, if not a general one. The International Union of Biological Sciences sponsored three summer conferences on theoretical biology at Bellaggio, Italy, in 1966–8. The chairman, C. H. Waddington, pointed out that in contrast to theoretical physics, 'theoretical biology can hardly be said to exist as yet ... There is little agreement as to what topics it should deal with or in what manner it should proceed.' In the hope of bringing out ideas to serve as unifying frameworks, Waddington cast his net wide, inviting geneticists and philosophers, specialists in nerve networks and computer design, experimental biologists, physicists, chemists and mathematicians – René Thom, E. C. Zeeman and David H. Fowler (also of Warwick).

The discussions at Bellaggio were intense and far-ranging, and the four volumes of papers by the participants contained a wealth of new ideas. Thom's essays, adapted from his book, dealt with his dynamic theory of morphogenesis, with the general role of topological models in biology and with applications of structuralism (as Thom interpreted it) to biology. Fowler, then engaged in translating Thom's book, proposed one of the elementary catastrophes as a qualitative model for phase transition in physics. Zeeman contributed a sketch of his topological brain modelling (written with Peter Buneman of Warwick) and more specific models, based on catastrophe theory, for two physiological processes: the heartbeat and the nerve impulse. This last was immediately provocative, because research on the nerve impulse – and a quantitative model for its biochemistry – had won a Nobel Prize in 1963 for Alan Hodgkin and Andrew Huxley. Now Zeeman, combining their data with his own mathematical scheme, advanced an alternate model implying different biochemical features that could be experimentally tested. He suggested that his scheme provided a better conceptual understanding by modelling 'the dynamics (which is relatively simple) rather than the biochemistry (which is relatively complicated)'.

The reactions of the other participants to these early presentations of catastrophe theory were mixed. Christopher Longuet-Higgins called the topological approach of Waddington and Thom 'an original and sophisticated approach to a wide class of problems, both evolutionary and morphogenetic. But', he asked, 'is its descriptive power equalled by its explanatory power?' (Waddington replied that the idea of canalized, stable processes had led him to discover a new genetic mechanism in evolution, and asked, 'What more can you demand of a piece of theoretical science?'). Biologist Brian Goodwin, whose experimental and theoretical work on temporal cycles in cells has been widely praised, wrote: 'The power of the topological approach is to be found in the generality of its analysis; it gives one both a high level of abstraction and analytical precision. This is amply illustrated in Thom's work. However, I believe that the qualitative insights provided by topology into the behaviour of dynamical systems must be combined with quantitative analysis.'

That was what Zeeman had attempted. Although the mathematical elegance of his models was admired, their scientific utility was not so clear. Jack Cowan, director of mathematical biology at the University of Chicago, later commented that although Zeeman's nerve-impulse model required fewer equations than the Hodgkin–Huxley model, it depended upon questionable *ad hoc* assumptions : 'It's a nice over-all picture, but I think it gives less rather than more insight into the chemical processes involved.'

THE DIVERGENCE OF THE THEORY

In 1976, looking back on the Bellaggio meetings, Goodwin recalled, 'They infected quite a number of people with the ideas of catastrophe theory. Of course, there were polar reactions for and against the ideas – but the infection "took" on both sides of the Atlantic.'

Other opportunities for the spread of the theory followed quickly. Thom wrote and lectured on it as an 'art of models', a way of generating and classifying analogies both within and across disciplines. He addressed himself to biologists, to linguistic and semantic specialists and to physicists, stressing that catastrophe theory's models were inherently qualitative and that in many, perhaps most, cases, they would be of more value in organizing existing data and suggesting new types of experiment than in prediction. Thom's book, *Structural Stability and Morphogenesis*, finally appeared in 1972, with a foreword by Waddington commending it as 'a very important contribution to the philosophy of science and to theoretical biology in particular'.

Zeeman concentrated on developing specific models that would attract the interest of researchers, at first in biology and then increasingly in the social sciences. To illustrate one of the elementary catastrophes he designed a 'catastrophe machine', an arrangement of cardboard and elastic that jumps suddenly from one position to another and back, although the movement that leads to the jump is smooth and continuous. With Carlos A. Isnard of IMPA in Brazil he presented a group of models for discontinuous behaviour in the social sciences – for example, a government's decision to go to war. In 1975 Zeeman collaborated with a Warwick statistician and three prison psychologists in developing a model for the sequence of events leading up to a 1972 riot in a British prison. They gathered and analysed data reflecting levels of tension and alienation among the prisoners, attempted to fit them quantitatively to an elementary catastrophe pattern, and suggested that the model could be tested for predictive accuracy if it were made part of an ongoing monitoring system. (Zeeman's prison model would later provoke the first flare-up of controversy over catastrophe theory's applications in the social sciences.)

*

By the time *Structural Stability and Morphogenesis* appeared in English, there was no longer any doubt about the increasing divergence in Thom's and Zeeman's views of how catastrophe theory should be used (although there was no lessening of their friendship and mutual respect). In an article that appeared in a student magazine of the Mathematics Institute in 1973, Thom wrote :

The catastrophe model is at the same time much less and much more than a scientific theory; one should consider it as a *language*, a method, which permits classification and systematization of given empirical data ... In fact, any phenomenon at all can be explained by a suitable model from catastrophe theory. And, as the English biologist Lewis Wolpert [another Bellaggio participant] remarked to me very pertinently, a theory that explains everything explains nothing. This simply shows that one should not expect from the model the same uses that one expects from a quantitative law of physics, or from an experimental trial ...

Thom also emphasized that the elementary catastrophes were only the first elements of a far more comprehensive theory, the development of which would require mathematical advances that he could only hope would be forthcoming.

Zeeman agreed that 'elementary catastrophe theory does not equal catastrophe theory. My emphasis on elementary catastrophe theory has been mainly because of its usefulness in applications.' But any theory that dealt with reality as well as mathematical abstractions, he contended, 'must face up to the classical scientific method of prediction, experiment and verification. I see no reason why [Thom's] theories should be sacrosanct on the grounds of being qualitative rather than quantitative. There are plenty of qualitative predictions in science, and plenty of quantitative experiments in which the quantities depend on the individual, but the quality is common to all individuals.' (For example, in the prison riot model, the quantitative levels of tension and alienation were different for each prisoner, but the qualitative behaviour – the outbreak of a riot – was a group phenomenon.)

The difference between the two men's views lies in what

Thom means by 'explain' and Zeeman means by 'verification'. For Thom, catastrophe theory's merit is that it makes our descriptions less arbitrary: it provides a common language for physical, biological and psychological processes. It avoids the dependence of quantitative theories on mysterious, irreducible forces. It is 'verified' only by its coherence, its mathematical power and its success in description. It has little to add to our understanding of those processes for which we possess clear quantitative theories – keeping in mind that such processes are only a small part of all that we observe. 'What I am offering', as Thom wrote in *Structural Stability and Morphogenesis*, 'is not a scientific theory, but a method; the first step in the construction of a model is to describe the dynamical models compatible with an empirically given morphology, and this is also the first step in understanding the phenomena under consideration ... We may hope that theoreticians will develop a quantitative model [for specific processes described by catastrophe theory] ... But this is only a hope.'

For Zeeman, on the other hand, qualitative modelling and quantitative theorizing are much more closely linked. He takes it for granted that catastrophe theory should permit qualitative prediction, suggesting that it may let the sociologist, for example, 'redesign his experiment with not only the aim of drawing a smooth curve to illustrate the trend, but also the aim of detecting those critical points where the curve ... may be discontinuous, and hence revealing the social morphology ...' Beyond that, he believes that it can be augmented with quantitative data so that its predictions can be tested against those of other theories.

The distinction between Thom's and Zeeman's views was lost in catastrophe theory's sudden emergence into the limelight in 1975 and 1976. Newspaper accounts in England led to a television programme about it on BBC's *Horizon*. An article in *New Scientist* was heralded by a cover showing the words CATASTROPHE THEORY in giant, cracking stone letters, like an advertisement for a Hollywood disaster film. *Newsweek* devoted two pages to the theory, stressing Zeeman's optimism about its applications and predictive value. Within a short time, the number of people who had at least heard of the theory increased a hundredfold, and a controversy over its merits sprang up. People who knew nothing of the theory's limitations spoke confidently of its revolutionary impact, and people who knew nothing of its power disparaged it as old facts dressed up in new, thoroughly unscientific analogies.

The controversy will be explored in Chapter 4; first, though, let us look at the models used in most applications of catastrophe theory so far – the elementary catastrophes. An understanding of their limitations, and their power, will put the arguments into perspective.

3

THE ELEMENTARY
CATASTROPHES

*'The reader may very well pause at this point and
ask what on earth we are talking about. How could
a graph possibly be like an overhanging cliff? And
why bother about three-dimensional graphs any-
way?'* – E. C. ZEEMAN

A CATASTROPHE, in the very broad sense Thom gives to the
word, is any discontinuous transition that occurs when a system can have more than one stable state, or can follow more
than one stable pathway of change. The catastrophe is the
'jump' from one state or pathway to another. In the landscape
imagined by Waddington, it could be represented as a passage
of an object from one basin to another, or as a flow of water
from one channel into another. The transition here is discontinuous not because there are no intervening states or pathways, but because none of them is stable: the passage from the
initial state or pathway to the final one is likely to be brief in
comparison to the time spent in stable states.

The elementary catastrophes are the seven simplest ways
for such a transition to occur. They can be illustrated by graphs
that show the stable states as sets of points – lines or surfaces –
in a 'behaviour space'. As long as the system 'occupies' one of
those points, its behaviour is continuous – but when it leaves
the line or surface, it is unstable and must return, sometimes at
a point far distant from the initial point. The graphs of the
seven elementary catastrophes depict seven topologically distinct arrangements of the points representing stable states.
Within certain limits, moreover, they are the *only possible*
arrangements. So they are, in a sense, archetypes: the most
basic models for many processes that are very different in

quantitative terms and in their internal workings. The graphs allow us to incorporate a great deal of information about causes and effects in a clear descriptive diagram. They are geometrically rich, with structural features that are not immediately apparent. Often, if a process is modelled with one of them because its behaviour corresponds to some features, we can then study the model to see what other, less obvious types of behaviour it suggests.

THE SLOPES OF STABILITY

When Laplace was working on celestial mechanics at the beginning of the eighteenth century, he developed a convenient mathematical shortcut to represent the action of gravitational force. This was the *potential*, a concept that summed up all the forces acting on an object in a single quantity. Instead of saying that the object changed its motion until no further forces acted upon it, one could say that it moved to a position of *minimum potential*. This enabled Laplace to bring to bear the techniques of the calculus of variations : finding the eventual position of the object meant finding a minimum solution for the equation of potential.

It has become customary since then to view many systems as governed by the tendency to seek a minimum of potential energy, although the energy may be of many different kinds. In a physical system, examples are the tendency of a stretched spring to contract, the tendency of two chemicals in a battery to react, or the tendency of a ball to roll downhill. The first is a mechanical potential, the second a chemical potential, the third a gravitational potential. Given the chance, all three potentials will decrease spontaneously, releasing energy as they do so. To increase the potential, conversely, requires that energy be put into the system by pulling on the spring, charging the battery or pushing the ball uphill.

Biology, social science and ordinary language also use the concept of potential. A plant grows upwards while its roots grow downwards because certain cells respond in specific ways to the gravitational potential; an animal eats until it is sated because of shifting chemical and electrical potentials in its blood and brain. To say that the plant's roots are geotropic (earth-seeking), or that the animal is hungry, is another way of saying the same thing. Psychologists and economists would find it impossible to theorize without assuming that there are drives, whether instinctive or learned, that individuals and groups seek to satisfy. Even the most resolute behaviourist, who firmly refuses to speculate about the psyche of a laboratory rat or a human being, recognizes that a slight stimulus can, in many cases, bring an energetic response; the difference in energy must come from somewhere. We say that we're 'tense' or 'under pressure', or that 'something's got to give'. These potentials are certainly far more complex and far more difficult to measure, but they are no less real.

The concept of potential is closely linked to that of *equilibrium*. When the spring has relaxed, or the battery has discharged, or the ball has rolled to the bottom of the slope, the potential is at a minimum and the physical system is in equilibrium. There are several kinds of equilibrium. A ball can be balanced on top of a hill, but the slightest push will start it rolling down : its equilibrium is unstable. If it is on a narrow ledge, a push in one direction will leave it there, but a push in the other direction will send it over the edge : its equilibrium is semi-stable. If it is at the bottom of a valley, it will resist a push in any direction : its equilibrium is stable.

In living systems, equilibrium is dynamic rather than static, because organisms and societies are always taking in and transforming energy. They tend to establish cycles in which no one state is stable, but the whole series of states resist disturbance like a spinning gyroscope. An example is the cycle of waking and sleeping, which can be upset by an unnatural change such as air travel across time zones, but (fortunately for the jet set) tends to re-establish itself. We take that cycle so much for granted that the persistence of one state – insomnia, for example, or coma – is a sure sign of some physiological or mental imbalance.

For a graphic representation of potential and equilibrium, look again at the roller-coaster curve of Figure 1, on page 26. There the y (vertical) axis represents levels of a potential – call it height, which is what it looks like, and which is equivalent to gravitational potential. The x (horizontal) axis represents some condition – call it the straight-line distance covered on the roller coaster – whose value determines the value of the potential. Now imagine that you can place a ball at any point on the curve. At all but four points the ball will immediately begin to roll. Those four points, the four places where the curve has neither upward nor downward slope, are equilibrium points. One is a 'ledge', one is a 'hilltop' and two are at the bottom of 'valleys'. Only the two minima are points of stable equilibrium; the point of inflection is semi-stable, and the local maximum is unstable.

This picture can be extended to three dimensions, giving us something more like a landscape, as in Figure 2. In this case the potential represented by y depends on two conditions, represented by x and z. If this were a real landscape, any combination of values for x and z (the latitude and longitude) would determine a unique elevation. This picture lets us visualize a dynamic equilibrium such as a ball rolling around and around the basin surrounding the local minimum. If it rolls without friction, it corresponds to what physicists call a 'conservative' system, one in which no energy is lost, and the cycle can continue indefinitely. If there is friction, the system is 'dissipative': unless energy is added to make up for its losses, the system will spiral down towards the minimum and eventually come to rest. (Nearly all physical systems, and all biological ones, are dissipative.) Notice that instead of a point of inflection, the surface in Figure 2 has a saddle point – a place which is at the crest of a ridge when approached along the z axis, but at the bottom of a trough when approached along the x axis. Thus, with a third dimension, a new qualitative feature becomes possible. The saddle point could not occur on a two-dimensional curve.

So far, we have looked at fixed graphs of potential. If they presented a complete picture, the world would be a dull place.

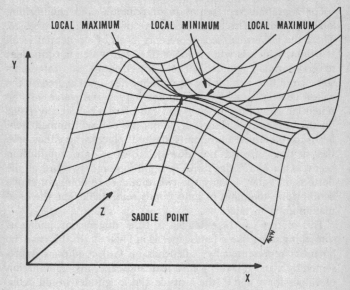

Figure 2. Singularities of a surface

Every system would find its way to a stable point or cycle, and that would be all. But the fixed picture is not complete, because the factors that determine potentials are not fixed. The landscape changes from moment to moment: 'every valley shall be exalted, and every mountain and hill shall be made low,' as the prophet Isaiah put it. And in many cases, a continuous change in the landscape can bring about a discontinuous change in the behaviour of systems.

OVER THE EDGE

Have you ever played with a stamped-metal 'clicker'? It's simply a disc with a moulded bulge in the centre. When you press down on the bulge it resists, then pops out on the other face of the disc. It makes a sound that is surprisingly loud, but hardly what you'd call catastrophic. Yet that's what a catastrophe is in Thom's theory : a sudden transition from one state of minimum potential, one stable equilibrium, to another.

Figure 3 shows the catastrophe in terms of the potential, which, in the clicker, takes the form of stresses in the metal. As you apply pressure, you alter the pattern of stress, so that the graph of potential changes its shape. (It's important to distinguish between the curvature of the metal and the curvature of the graph, which are mathematically related but not the same.) First a point of inflection appears on the curve, then a new local maximum and minimum. If you don't press the clicker hard enough, it will return to its original form as soon as you let up. As long as the original minimum of potential lasts, no matter how shallow it may be, the ball on the curve goes nowhere. But

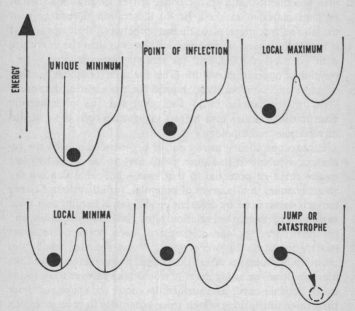

Figure 3. A simple catastrophe: A sudden change in potential energy

47

if you do press hard enough, the 'valley' turns into a 'ledge', and when even that disappears catastrophe is unavoidable. The system has to jump : the bulge of the clicker pops out on the other side, the symbolic ball rolls downhill to the only remaining minimum.

At the third, fourth and fifth stages shown in Figure 3, there is a third possible equilibrium point, the unstable maximum. If you are very careful, you can force the clicker into a flat configuration, but it's unlikely to stay flat for long. The slightest touch will shift the local maximum of the potential curve 'out from under' the ball, and . . . CLICK !

This kind of discontinuity can be found in many processes. A light switch shows similar behaviour, its continuous range of motion producing one of two discontinuous stable states. Zeeman points out that a plastic coffee stirrer, or any long strip of resilient material, makes a handy illustration. Spreading your thumb and forefinger, place the stirrer between their tips so that it is slightly bowed by compression. Press against the curvature with your other hand, and the strip buckles and jumps to a position of opposite curvature. (This example is familiar to every engineer as 'Euler buckling', named for the eighteenth-century Swiss mathematician who first analysed the phenomenon quantitatively. Euler also helped inaugurate both the calculus of variations and topology.)

Catastrophe theory grows out of a generalization of the behaviour symbolized in Figure 3. It's easy to see what happens to the curve of potential in that case – but what can we say about changes in *all* curves of potential, for *all* systems? Every curve is determined by different conditions, according to a wide variety of mathematical relationships. The curve's position, and the relative height and steepness of each portion, are never exactly the same for two systems. Worse yet, the possibilities multiply in processes where the potential depends on more than one condition, as in Figure 2. It would take forever to examine every possible curve and surface, let alone the shapes of more than three dimensions which trace potentials in more complex processes.

But perhaps it isn't necessary to examine every one. In Figure 3 the clicker's behaviour tells us that a shallow minimum is often as good as a deep one. No jump occurs until the minimum vanishes altogether and the intervening maximum is gone. So the position, the height and the steepness of the curve are irrelevant as far as catastrophe is concerned. All we need to know is the curve's *qualitative* shape – and that changes only when an equilibrium point is created or destroyed.

This is a topological approach to the question, and it simplifies matters by letting us treat whole classes of curves at once. In Figures 4 and 5 we can see how it works. Figure 4 shows a set of curves obtained by graphing a relationship between quantities x and y (curve b), then 'tinkering' with it by adding and subtracting an arbitrary amount of x in the equation (curves a and c respectively). The curve moves up and down, but keeps its qualitative feature: a single, unique minimum. Figure 5 shows the curves obtained by doing the same thing to a different equation. The original curve, b, has a point of inflection; curve a has no equilibrium points at all; curve c has a local maximum and a local minimum. So the equation graphed in Figure 5, *and all equations of the same general type*, are candidates for catastrophe. Any system whose potential follows that type of equation will have a stable state under some conditions, then as the conditions change only a semi-stable state, then none at all. This is the simplest type of catastrophe. The next type corresponds to any curve of potential which has first one minimum, then two local minima with an unstable local maximum between, then only one minimum again. That is the type of catastrophe symbolized by Figure 3.

It was this process of topological abstraction, carried a great deal farther, that made possible Thom's 'classification theorem' of 1965. He had already developed his ideas concerning structural stability in nature, and he was seeking topological models compatible with those ideas. The models must depict both continuous and discontinuous change, and they must themselves be stable – that is, they must retain their qualitative structure in spite of small quantitative variations. There is a subtle paradox

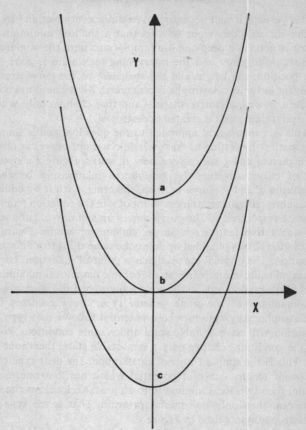

Figure 4. A qualitatively stable set of curves

here : each model summarizes the appearance and disappearance of stability, but it does so in a stable way. This is possible, as Thom discovered, because the equilibrium points for general classes of equations can be represented as unfoldings of topological singularities, and because for each of the seven simplest

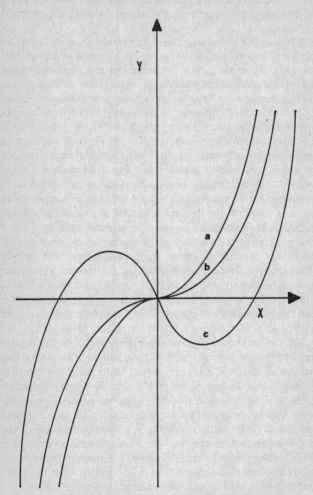

Figure 5. A qualitatively unstable set of curves

singularities, there is only one stable unfolding : others are possible, but they 'collapse' into the stable form at the slightest disturbance.

It is impossible to present the classification theorem itself, let alone its proof, except in the technical language of differential topology. But the theorem's consequences can be stated in the language we have already used : *In any system governed by a potential, and in which the system's behaviour is determined by no more than four different factors, only seven qualitatively different types of discontinuity are possible.* In other words, while there are an infinite number of ways for such a system to change continuously (staying at or near equilibrium), there are only seven structurally stable ways for it to change discontinuously (passing through non-equilibrium states). Other ways are conceivable, but unstable; they are unlikely to happen more than once, and lack the 'recurrent identifiable elements' needed to establish themselves in language or scientific theory.

The qualitative type of any stable discontinuity does *not* depend on the specific nature of the potential involved, merely on its existence. It does *not* depend on the specific conditions regulating behaviour, merely on their number. It does *not* depend on the specific quantitative, cause-and-effect relationship between the conditions and the resultant behaviour, merely on the empirical fact that such a relationship exists. Now we can see how the elementary catastrophes are comparable to the regular forms of classical geometry. Just as we can say that any three-dimensional object, *if* it is regular (i.e., all its faces are identical polygons), must be one of the five solids, so Thom's classification theorem asserts that any discontinuous process whose behaviour can be described by a graph in as many as six dimensions, *if* it is structurally stable, must correspond to one of the seven elementary catastrophes. The first, geometric assertion is true no matter what the size or material of the object involved; Thom's topological assertion is true no matter what the magnitude, internal workings or nature of the process involved.

To put it very simply, in a wide range of situations – physical, biological, even psychological – where experience tells us that 'something's got to give' (i.e., there is a potential and a possible discontinuity), the classification theorem indicates that there are only seven fundamentally different ways it can happen.

THE SURPRISING SEVEN

To represent these families of behaviour, we need a new kind of graph, quite different from those in Figures 1 through 5. The new graph must have one dimension, or axis, for each *control factor* that determines a system's behaviour. It must have an additional axis or two to represent the behaviour itself. In the space defined by these dimensions, every possible equilibrium state of a system is represented by a single point, and the points form a smooth line or surface. A continuous change in behaviour appears as a movement *within* the line or surface; a discontinuous change appears as a movement that *leaves* the line or surface. The simplest elementary catastrophe, the fold, has only one control axis and one behaviour axis, and is thus two-dimensional. The most complex, the parabolic umbilic, has four control axes and two behaviour axes, and is thus six-dimensional.

The following table summarizes the elementary catastrophes: The names of the four catastrophes in the first column were suggested by visual features of the graphs that depict them,

Number of control factors	One behaviour axis	Two behaviour axes
1	fold	—
2	cusp	—
3	swallowtail	hyperbolic umbilic elliptic umbilic
4	butterfly	parabolic umbilic

while those in the second column – which are harder to visualize – have names that are starkly mathematical. The gift of visualization can be developed in surprising ways: the name of the swallowtail, *la queue-d'aronde*, for example, was suggested by the blind French mathematician Bernard Morin.

The *fold catastrophe* graph represents the behaviour of all systems dependent on one single varying condition, or control factor. The arrangement of possible equilibrium states is shown in Figure 6. Notice that the potential governing the system does

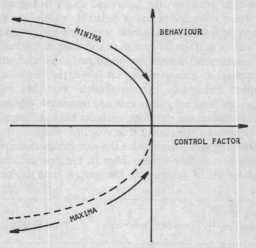

Figure 6. The fold catastrophe graph

not appear, as it did in Figures 1 through 5; instead, it is implicit in the fact that the graph shows only the three types of equilibrium states: maxima, minima and a point of inflection where the curve meets the behaviour axis. Any pair of values for x (control) and y (behaviour), any point of the plane, stands for a single combination of the control factor and the behaviour. It is the potential that 'pulls' the system towards stable equilibrium points, those on the solid curve. A catastrophe occurs whenever the point crosses the control axis.

The qualitative type of behaviour characterized by this graph is simple. For one range of values of the control factor, there are both maxima and minima and the system tends to move spontaneously to the minimum. At a critical or threshold value of the control factor, the maxima and minima fuse in a point of inflection. Here there is only one semi-stable state, the point where the x and y axes intersect. Beyond the critical value, there are no equilibrium states at all. This means that a system in this condition is completely unstable. Potentials that accompany this type of behaviour are like those graphed in Figure 5.

The fold catastrophe has little to tell us, since there are only a few things that can happen in such a system, all of them obvious. The system can move to a state of minimum potential if conditions allow one to exist; it can be balanced at the point of inflection; or it can be essentially unstable, having no available minimum. An example of such a system is a rubber band, in which the control factor is the force applied to stretch it and the behaviour of its tension. Up to a critical level of force, the rubber band is taut and straight – that is, it minimizes the tension by being as short as it can. Beyond that critical level, the rubber band snaps, and there is no longer any tension to measure. The broken pieces may lie in any curve at all; no position is more stable than any other.

The *cusp catastrophe* occurs in systems whose behaviour depends on two control factors. Its graph (Figure 7) is three-dimensional, a curved surface with a pleat. Again, every point of the surface represents an equilibrium state. All the points on the underside of the pleat are unstable maxima. All the points along the fold line, which forms the 'lip' on the pleat, are semi-stable points of inflection. All the rest of the points are stable minima.

For certain combinations of values of the control factors, there are two possible stable states, one on the upper surface of the pleat and one on the lower surface beneath the pleat.

The behaviour of the system under these conditions is called 'bimodal', meaning that the same conditions permit either of two stable states. (There is a third possibility, the unstable maximum on the underside of the pleat, but this is generally inaccessible: if the system occupies this state, any disturbance will force it to the stable point above or below.)

Now, imagine the conditions changing with time, so that the system's behaviour changes. All smooth changes can be visualized as points moving along the surface. Take two points close

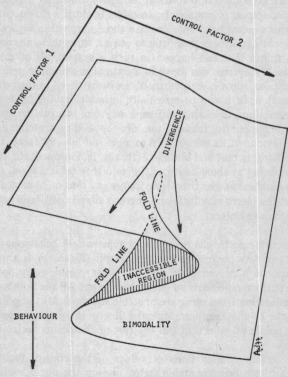

Figure 7. The cusp catastrophe graph

to each other at the far edge of the surface (that nearest the axis of control factor 2) in Figure 7. They represent systems at the same value of control factor 1, but slightly different values of control factor 2; they are side by side, one a little higher than the other. If the value of control factor 1 then increases, the points move forwards towards the front of the surface, tracing parallel paths. If both of them pass on the same side of the pleat, the behaviour of the two systems remains similar – but if one travels on to the upper surface of the pleat, while its neighbour travels on to the surface beneath the pleat, then the behaviour of the systems is *divergent*. They begin to close together, they undergo the same change of conditions, but at the end of that change, they are far apart in behaviour. Which path a point takes in this case depends on the precise value of control factor 2 as the moving point passes the beginning of the pleat.

Even the divergent paths are still smooth (non-catastrophic) changes in behaviour. But the cusp catastrophe graph also suggests the possibility of *discontinuous* changes, those which occur when a point moving to the left or right reaches the lip of the pleat. Figure 8 shows the situation: the system can pass smoothly from *a* to *c* and back, *a* to *b* and back, *b* to *e* and back. But if the system is at *c* and control factor 2 is increased, the point reaches *d* ... and there isn't anywhere else to go. What was a stable minimum has turned into a point of inflection, and any further increase in control factor 2 obliges the system to 'jump' to the only stable minimum left, the one at *e*. It passes as quickly as possible through the non-equilibrium states; the transition is a catastrophe. A similar jump occurs if a system at *e* is altered by a decrease in control factor 2: it moves to *f*, then has to jump catastrophically to *c*. Both jumps are equivalent to the discontinuity whose potential curves were shown in Figure 3. The cusp catastrophe is the model which describes the behaviour of the metal clicker.

Figure 8 shows that it is possible to get from *c* to *e*, for example, either smoothly or via a catastrophe. Which will occur in any particular case depends on the sequence and the degree of the changes in the control factors. In an experiment, we

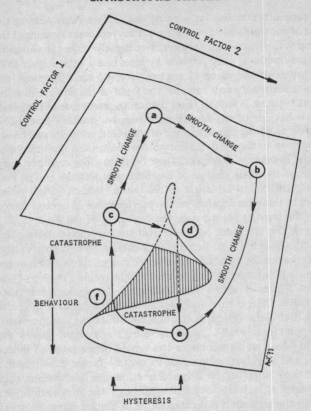

Figure 8. *Continuous and discontinuous changes shown as paths on the cusp catastrophe graph*

would need to have both factors subject to our control in order to have a choice between continuous and discontinuous paths. If a system is at point *c* and control factor 2 alternately increases and decreases by a suitable amount, the result is a cycle of behaviour with two smooth portions linked by catastrophes. Such a cycle is called *hysteresis*, and it is found in many dynamic systems, from electrical circuits to manic-depressive psychoses.

So the cusp catastrophe model adds a number of features not seen in the fold model: bimodality, divergence, two sets of catastrophic jumps, smooth or sudden passages between the same initial and final states, and hysteresis. The model is valuable because so many processes in the real world appear to have this cluster of behaviour types. As a result, the cusp is the model most frequently used in qualitative applications. Take the cycle of waking and sleeping, for example. Earlier in this chapter, it was treated as a continuous cycle, but on closer examination, it appears rather like a hysteresis cycle. Our state of consciousness changes smoothly from moment to moment while we are awake or asleep, but the transition between the two levels of awareness is relatively smooth in some cases, sudden and discontinuous in others. Is there some rhythmic increase and decrease in a physiological 'control factor 2', perhaps in nerve impulses or levels of a neurotransmitter chemical in the lower brain, that triggers drowsiness or alertness? When we are half-awake, a slight difference in the intensity of an external stimulus can make the difference between drifting off to sleep and becoming alert: our response is divergent.

For another, more concrete example, consider the transition of water between its liquid and gaseous states. This is usually a discontinuous change, though it doesn't have to be. At a sufficiently high temperature and pressure (374°C and 218 atmospheres), water changes from liquid to vapour and back without boiling or condensation. Figure 9 shows water's behaviour plotted on the cusp catastrophe surface, with temperature and pressure as the control factors and the water's density as the behaviour. The path from a to d can be smooth at sufficiently high values of the control factors, or it can be discontinuous, as it is under what we consider normal conditions (100° C and 1 atmosphere).

Experience tells us that under normal conditions, water boils and condenses at the same temperature. This is not consistent with the behaviour we have plotted so far on the cusp catastrophe surface; in Figure 9 the behaviour point makes its

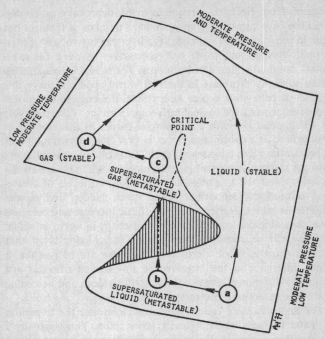

Figure 9. A cusp catastrophe model of boiling and condensation (NOTE: *The control axes are oriented at 45° to their position in the other figures.*)

transition (*b–c*) before reaching the lip of the pleat. The reason is that boiling and condensation are not really single transitions, but 'averages' of the transitions for millions of molecules. For this kind of phenomenon the catastrophe follows what is called 'Maxwell's rule', after the nineteenth-century physicist James Clerk Maxwell. In other cases, it follows the 'delay rule', meaning that the behaviour point stays in the original stable minimum as long as it can. Previously, we have implicitly used the delay rule in describing behaviour on the cusp. Most of the processes discussed in the following chapters obey that rule. (By the way, water can be made to follow the delay rule, too. If heating and cooling are carried out very carefully, water can be 'superheated' and water vapor can be 'supercooled'; the eventual transitions in each case *do* occur at different temperatures.)

One final note on the mathematical structure of the cusp catastrophe: if you can visualize a cross-sectional 'cut' in the surface, parallel to the axis of control factor 2, what you see depends on where the cut is made. If it transects the rear portion of the surface, the cut shows a smoothly decreasing curve, steepest in its centre portion. If it passes through the beginning of the pleat, the cut shows a curve that becomes momentarily vertical at the centre, then flattens out again. If the cut passes through the pleat, it shows an S-shaped curve that doubles back on itself, looking like two interlocked fold curves. In fact, it *is* two interlocked fold curves. Each of the elementary catastrophes is built up out of those of lower dimension. The cusp catastrophe contains two fold catastrophes linked at a topological singularity, the point where the pleat begins. Similarly, the swallowtail contains a cusp catastrophe and a fold catastrophe linked at a more complex singularity ... and so on.

The *swallowtail catastrophe* can be used to model processes in systems where behaviour depends on three control factors. Its graph is four-dimensional, so that even a three-dimensional model (let alone a drawing!) is inadequate. But three-dimensional 'slices' of the graph can be obtained by fixing the value of one of the control factors, as shown in Figure 10. In part of its range (10a), the surface is simply a folded sheet. In another part (10b), it develops an internal kink that resembles the outline of a bird's tail. Outside the kink, the swallowtail has one stable state for each set of conditions. Inside, it has two: a straight line through the kink would pass through the surface four times, twice at maxima and twice at minima. In a swallowtail model, catastrophe occurs when a system leaves the surface – whether it is to another layer of the surface or to a position not on the surface.

The swallowtail catastrophe is not particularly useful as a

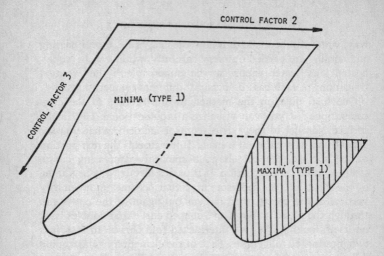

CONTROL FACTOR 1 FIXED IN VALUE

Figure 10a. *A three-dimensional view of the swallowtail catastrophe graph*

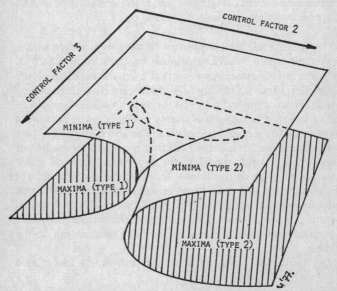

Figure 10b. *Another view of the same graph*

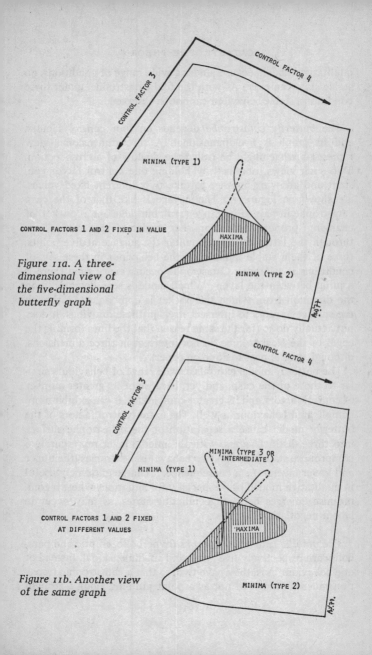

CONTROL FACTOR 3

CONTROL FACTOR 4

MINIMA (TYPE 1)

CONTROL FACTORS 1 AND 2 FIXED IN VALUE

MAXIMA

MINIMA (TYPE 2)

Figure 11a. A three-dimensional view of the five-dimensional butterfly graph

CONTROL FACTOR 3

CONTROL FACTOR 4

MINIMA (TYPE 1)

MINIMA (TYPE 3 OR INTERMEDIATE)

CONTROL FACTORS 1 AND 2 FIXED AT DIFFERENT VALUES

MAXIMA

MINIMA (TYPE 2)

Figure 11b. Another view of the same graph

qualitative model because, under a wide range of conditions, no stable state can exist. As with the fold catastrophe, under these conditions, stable behaviour cannot be observed.

The *butterfly catastrophe* depends on four control factors, and its graph is five-dimensional. A three-dimensional view represents what might be called the shadow of a cross-section. Two such views, obtained by holding one control factor constant and allowing another to take two different fixed values, are shown in Figure 11. The surface is like that of the cusp catastrophe in part of its range (11a), but develops a 'pocket' of changing proportions in another region (11b). Some lines through the latter region encounter the surface at five points, three of them stable minima, so the behaviour in these sets of conditions is trimodal. Catastrophe occurs every time there is a jump between the layers. Which mode is seen depends upon the direction from which the pocket is approached. Although the surface appears to intersect itself in these drawings, it does not actually do so (for the same reason that the lines forming the edges of the Necker cube do not intersect in three dimensions, although they appear to do so on paper).

The butterfly model exhibits a wide range of behaviours similar to those of the cusp, and yet, because of its greater number of control factors, and its greater complexity, it can exhibit more complicated behaviour as well. The extra control factors of the butterfly model cause a separation of the catastrophe surface into three distinct layers, with the middle layer representing a compromise state between two behavioural extremes (the upper and lower layers). As a consequence, it can be extremely useful in qualitative modelling – especially for situations where a compromise emerges between conflicting states, as may occur in labour negotiations.

The *umbilic catastrophe graphs* (hyperbolic, elliptic and parabolic) are respectively five-, five- and six-dimensional. Instead of one behaviour axis, they have two, so that a catastrophic transition must be imagined not as a point jumping along a straight line (as in the cusp catastrophe graph) but as a line jumping across a plane. Obviously, these three types of umbilic catastrophe are 'elementary' only in a technical sense. Their geometry is very rich : consider how many features of behaviour are possible with just the cusp, and extend that variety into several additional dimensions. Even the experts can lose their way in this complexity, as Thom discovered when his early identification of the hyperbolic umbilic catastrophe with the shape of a breaking wave turned out to be mistaken. It has proved possible to study these complex forms by programming a computer to draw planar projections of them for various combinations of values of the control factors. Samples of these projections are shown in Figure 12 (hyperbolic umbilic), Figure 13 (elliptic umbilic) and Figure 14 (parabolic umbilic). As with the other models, catastrophe occurs in the umbilic models every time the system leaves the surface.

The rich geometry of the umbilic catastrophes has suggested their usefulness as qualitative models for complex physical phenomena, such as those associated with geometric optics, engineering design and fluid dynamics. However, because of their extreme complexity, they do not immediately lend themselves to the relatively 'rough' modelling typical of the social sciences.

These seven forms are the products of what mathematician Ian Stewart calls Thom's 'wonderful theorem'. Since 1965 the theorem has been extended to describe systems with five control factors, thus adding another four catastrophes, even more complex than the original seven. For catastrophes with more than five control factors, there is an infinite number of singularities without unique unfoldings. When this occurs, it is no longer possible to distinguish among the possible catastrophe surfaces.

There are three important points to remember in using the elementary catastrophe graphs to model qualitative processes. First, they have *no scale*. Although we can say that the value of a given control factor increases in one direction and decreases in the opposite direction, we cannot say how rapidly it does so. In fact, the rate of increase or decrease need not be constant,

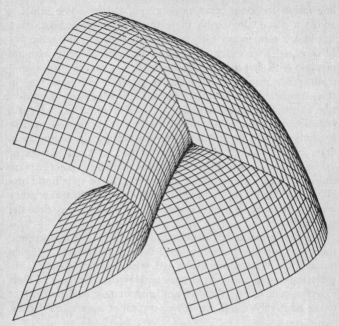

Figure 12. A computer-drawn projection of the hyberbolic umbilic catastrophe graph

so it is possible that nearly all of the cusp catastrophe surface, for example, is 'squeezed' into a narrow range of values for one of the control factors, while the remainder of the surface is 'stretched' over a much wider range. What this means is that a catastrophe graph in itself gives us no quantitative information. In order to identify any point on it with a specific level of behaviour, it is necessary to fit the surface to empirical measurements.

Second, these graphs show the *canonical* shape of each catastrophe surface; that is, they are basic or standard forms. In fact, by the classification theorem, any surface in up to six

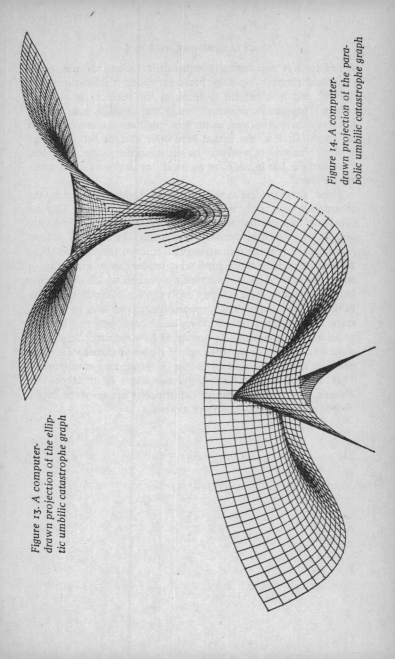

CATASTROPHE THEORY

Figure 13. A computer-drawn projection of the elliptic umbilic catastrophe graph

Figure 14. A computer-drawn projection of the parabolic umbilic catastrophe graph

dimensions is topologically equivalent to one of the seven canonical shapes. Topology has been called 'rubber-sheet geometry', and it is helpful to think of the surfaces as made of rubber. They can be deformed in any way at all, as long as they are not torn and as long as no new qualitative features appear on them. That is, the actual behaviour surface for a process with two control factors may be any distortion of the canonical cusp surface, but it can have no local singularity more complex than a pleat.

Third, these models in themselves are highly *idealized* by the assumption that only a single potential is involved. This assumption implies that an entire process can be modelled with one single type of elementary catastrophe. In complex natural processes, such as the formation of an embryo, there are multiple potentials, and there are conflicting tendencies to maximize or minimize each of them. As a result, a process which shows cusp behaviour at one moment may evolve into a butterfly or an umbilic. The elementary catastrophes themselves are, in a sense, static. Much of *Structural Stability and Morphogenesis* is devoted to an informal, intuitive discussion of how catastrophes of high dimension, although not themselves visible or classifiable, could generate a stable or unstable series of elementary catastrophes. That discussion goes far beyond the scope of this book; in Thom's view, the elementary catastrophes are only the beginning of such a 'general theory of models'.

4

THE CONTROVERSY

> 'We ... are excited about the prospect of new applications of mathematics, and concerned that many will be disenchanted with all of modern mathematics when they discover, as we have, that catastrophe theory is a blind alley' – R. S. ZAHLER AND H. J. SUSSMANN

> 'Some of my readers may agree with me that the degree of interest or importance of a mathematical or physical idea rarely corresponds to its formal rigor; the rest should perhaps wait a few years ...' – RENÉ THOM

THE more profoundly original an idea is, the longer it takes to transform our way of thought, and the harder it is to assess on first appearance. For example, Newton's theory of objects attracting each other at a distance had to overcome the opposition of the followers of Descartes, who had imagined space as filled with more or less tangible matter, and force as the effect of whirling vortices in that matter. 'Hence at that time', wrote physicist Philipp Frank in 1949, 'Newton's followers ... extolled his teachings as "mathematical" and "spiritual" in contrast to materialism.' Who could have imagined in, say, 1700 that Newtonian thinking would come to be almost synonymous with materialism?

Today, Newton's principles seem purest common sense. But as Frank's student, nuclear physicist Jeremy Bernstein, reminds us, 'The idea that the theories of the past are more intuitive and more solid is based on an optical illusion induced by the passage of time. Anyone, even a contemporary scientist, who studies Newton's *Principia*, with its fantastic array of geometric arguments and quasi-theological presuppositions, acquires a better notion of just how difficult and just how non-intuitive it was.' Perhaps the surest sign of a truly revolutionary idea is its ability, revealed over generations or even centuries, to make what once seemed abstract into the normal, the everyday, the intuitively obvious.

As we have seen, the ideas of catastrophe theory were beginning to circulate in the late 1960s, but the theory's broad scope and ambition did not become clear until the publication of *Structural Stability and Morphogenesis* in 1972. In three reviews that appeared in 1973, the reviewers – a mathematician, a biologist and a physicist – showed remarkably similar reactions. All of them noted Thom's extraordinary imaginative reach, all had reservations about the applications sketched out for the theory, and all were convinced of its long-range importance.

Both Newton's *Principia* and Thom's book, observed physicist C. W. Kilmister, of King's College, London, 'lay out a new conceptual framework for the understanding of nature, and equally both go on to unbounded speculation ... from Thom's will come an elaboration which will allow our children to use more precisely the concept of catastrophe that he introduces. Almost all of the speculation in the *Principia* has turned out to be hopelessly wrong, although it would be hard to say to what extent this prevented its being useful at the time. So the criterion for Thom's speculative applications must come from their stimulating character, without trying to prejudge their correctness.'

Brian Goodwin, a biological experimenter and theorist who had attended the Bellaggio conferences, wrote that 'practically every specific biological process which is described and analysed by the theory advanced in this book is more thoroughly and adequately accounted for at the moment' by models at specific levels (biochemical, physiological, genetic, etc.) than by Thom's more general models. Those he found 'full of improvisation and rich in suggestion'. Yet, he thought the book was 'remarkable ... for the breadth of the inquiry undertaken and for the unity and coherence of vision which are achieved', and added that, despite its shortcomings, it gave him 'a sense of liberation and enlightenment akin to what I imagine [sixteenth-century] astronomers must have felt when offered Copernican heliocentric [sun-centred] geometry ... Even if this biological model fails, the sustained inspiration and the vast scope of the book put it firmly into the best tradition of natural philosophy, the search for a rigorous and meaningful synthesis ...'

Reviewing the book for the American Mathematical Society *Bulletin*, John Guckenheimer, an expert in dynamic theory and topology, warned his audience at the beginning that it was not a work of mathematics, and that to apply the usual technical criteria for rigorous proofs would be a mistake, since Thom had made no pretence of having tried to meet these standards. 'Instead of insisting that Thom's style conform to prevailing norms,' Guckenheimer wrote, 'we should applaud him for sharing his wonderful imagination with us.' Guckenheimer found Thom overly optimistic in some of his mathematical and physical assumptions, and vague in linking the assumptions to the applications, but, like Kilmister and Goodwin, was enthusiastic about the general significance of the theory. He concluded his review by suggesting that Thom might even be too cautious about the impact of his mathematics on biology. 'Thom is pessimistic that this gap from singularity theory to experimental models can be bridged, but I am not so sure ... In the past, understanding Thom has been a rewarding experience. There is much for each of us to discover in this marvelous book.'

From the reaction to *Structural Stability and Morphogenesis*, one would assume that any comprehensive judgement of catastrophe theory would require many years, even generations, to develop. But the theory's passage from pure mathematics to detailed applications has been so swift, especially in Zeeman's hands, that a strong backlash has begun. The most articulate spokesman of this reaction is mathematician Hector J. Sussmann of Rutgers. Sussmann first studied catastrophe theory to present it to a Rutgers seminar, and his 1975 paper on its mathematical aspects was an admirably clear and well-informed summary. 'This author takes no position on the hotly debated question as to whether catastrophe theory really has important applications,' he wrote then.

By 1976 Sussman was taking a position – and heating up the debate. In a paper presented to the Philosophy of Science Association, he dissected three of Zeeman's models – for stock market crashes, prison riots and public opinion's effect on military policy – finding them to be 'vaguely formulated, to be based on false hypotheses, [and] to lead to few non-trivial predictions. Moreover, most of these predictions do not agree with reality.' He began work on a more extended attack in collaboration with Raphael S. Zahler of Yale. Months before publication, it was summarized by Gina Kolata of *Science* under the headline: CATASTROPHE THEORY: THE EMPEROR HAS NO CLOTHES. Sussmann was in demand as a lecturer, she reported, and there was a growing audience for his vigorous polemics. 'Sussmann convinces everyone,' applied mathematician Joseph Keller told her. 'Even people who supported the applications of catastrophe theory walk out of his lecture saying, "How could anyone believe that?"' In a subsequent issue of *Science*, a number of correspondents defended the theory.

Zahler and Sussmann presented the main points of their long critique in a late 1977 issue of *Nature* (the British counterpart of *Science*), where Goodwin's review and several articles on catastrophe theory in engineering and physics had appeared. They concluded that 'the claims made for the theory are greatly exaggerated and its accomplishments, at least in the biological and social sciences, are insignificant ... Catastrophe theory is one of many attempts that have been made to deduce the world by thought alone ... an appealing dream for mathematicians, but a dream that cannot come true.'

For scientists following these journals, the picture was confusing enough. For laymen encountering the theory only at third or fourth hand, it was bewildering. A news magazine told them in 1976 that catastrophe theory had been 'hailed as an "intellectual revolution" in mathematics, the most important development since calculus', and was being used to design bridges; by 1977 another magazine told them of 'the death of a theory', and expressed satisfaction that 'we laymen, whose mathematical qualifications reside somewhere on the wrong side of a balanced checkbook, can happily dismiss catastrophe theory as something else we need not feel guilty about not understanding a single word of'.

Both statements are misleading, or worse. The claim that the theory is an 'intellectual revolution' makes no sense for mathematics, which advances not by displacing old ideas but by broadening, deepening and combining them as Poincaré did. And while it is natural to compare catastrophe theory to calculus (for it was developed in part to balance calculus's emphasis on quantity and continuity), its 'importance' – its long-term impact on science – will not be clear for a long time. Calculus itself took a long time to become the foremost tool of mathematical science, and its conceptual foundations were not made rigorous until it had been in use for two centuries. As for technological tasks such as bridge-building, catastrophe theory's contribution is to the design of experiments on structural components rather than to the engineer on the job.

The second statement, which discounts catastrophe theory altogether, misses the mark even more widely, for it mistakes the theory's growing pains for mortal injury. It is reminiscent of the newspaper editorial in the 1920s that chided Robert Goddard, the pioneer rocket scientist, for not seeing that his designs would be useless in outer space, where there was no air for the rocket blast to push against.

In fact, the controversy surrounding catastrophe theory is easier to understand than the theory itself, once we realize that it is actually four arguments : one over the theory's foundations in mathematics and natural philosophy; one over the assumptions needed to apply it; one over the details of specific applications; and one over the attitudes, the style, even the intellectual honesty, of the theory's advocates and opponents. The argument over specific applications is best understood in context, and will be discussed in Chapters 5 through 9. The rest of this chapter will examine the other three areas of debate.

GLOBAL MISGIVINGS

Mathematics has two cutting edges : one in its formal abstractions, the pure manipulation of ideas, and one in its applications to the real world. For Thom, they are two edges of a single tool. He has faith in what physicist Eugene Wigner called 'the unreasonable effectiveness of mathematics in the natural sciences', and indeed finds it quite reasonable, given his views on the topological correspondence between the structure of the world and the structure of ideas.

But the application of such a tool, especially when the mathematics involved is new, is a matter of creative intuition and even personal taste. Thom, for example, though polite and reserved by nature, is said to have once interrupted a colleague's description of a new line of thought with, 'I don't like that conjecture at all ... and even if it's right, it's in the wrong spirit!' There are mathematicians who react with a similar instinct, perhaps even prejudice, to the conjectures with which Thom developed catastrophe theory from its mathematical origins in the study of singularities.

For one thing, they say, the classification of the elementary catastrophes depends on what is called 'local' analysis of topological properties – in other words, analysis that describes only the immediate neighbourhood of the singularity. But the classification theorem does not prove that a system's total range, its 'global' behaviour, is like its behaviour in that neighbourhood. The canonical drawings of Chapter 3 might represent only a small portion of global behaviour. Since the topological approach provides no scale, it requires an act of faith to identify a mathematical jump on the catastrophe surface with an observed discontinuity in nature. The justification of that faith, and the integration of the elementary catastrophes into the global scheme Thom outlines in his book, will require further advances in mathematics – perhaps by Thom, perhaps by other singularity theorists. There are conceptual obstacles, not merely technical difficulties, in the way of those advances.

One mathematician closely identified with global analysis is

Steven Smale, himself a Fields medallist, who takes a very critical view of catastrophe theory. 'It is more a philosophy than mathematics,' he says, 'and even as a philosophy it doesn't explain the real world . . . as mathematics, it brings together two of the most basic ideas in modern math : the study of dynamic systems and the study of the singularities of maps. Together, they cover a very wide area – but catastrophe theory brings them together in an arbitrary and constrained way.' For Thom, of course, the goal is not explanation but description, and the way in which he has combined mathematical ideas is a natural consequence of his assumption of structural stability.

Other mathematicians object to the claim by Zeeman that catastrophe theory is the first or the best way to approach discontinuity in nature. There are highly developed techniques for that already, they say, pointing to the mathematics of quantum theory, to the theory of shock waves, and especially to the very active field of bifurcation theory. Bifurcation theory, like catastrophe theory, can be traced back to Poincaré; it is the study of equations whose curves branch or 'bifurcate' at a critical point, so that two or more values of y are possible for a single value of x. Look back at Figure 7 on page 56 above, and vizualize a cross-sectional cut parallel to the axis of control factor 1, passing directly through the singularity and the centre of the pleat. The cut shows a single line up to a critical value of control factor 1, then two divergent branches. In fact, there is a close correspondence between catastrophe theory and bifurcation theory, and in many cases their mathematics are equivalent or overlapping. Bifurcation theory, though, is much more analytic in spirit.

At a New York Academy of Sciences conference on bifurcation theory and a wide range of its applications in late 1977, the effects of the controversy over catastrophe theory were evident. Some of the participants used the terminology and graphic symbolism of both theories, while others took pains to distance themselves and their work from catastrophe theory. It is quite possible that in the long run one of the two theories will be absorbed by the other, or that both will become part of some wider theory as yet unnamed; that prospect may have heightened the tension. Martin Golubitsky, a mathematician who has worked on applications of both theories in physics and engineering, began his talk with a slide showing a *New Yorker* cartoon by Fisher, in which a Viking ship with a fierce dragon prow was being approached by another ship, this one with a smiling rabbit as figurehead. On the sail of the Viking ship, Golubitsky had written HARD ANALYSIS; from the mast of the other flew a banner lettered CATASTROPHE THEORY. Golubitsky read the caption: the Viking captain saying, 'Careful, men, I don't like the looks of this!' There was sustained laughter in the audience.

Bifurcation theory also poses a challenge to Thom's principle of structural stability. In recent years, bifurcation theorists have made great progress in analysing equations whose curves branch again and again, until there may be infinitely many possible solutions under a given set of conditions. Many researchers believe that such equations may describe phenomena such as animal populations that fluctuate wildly from year to year (even under steady conditions), or the onset of turbulence in a smoothly flowing fluid. In some cases, these systems settle into a periodic cycle, but in others they wander without any pattern from one mode of behaviour to another.

'There are very reasonable models for occurrences in the real world that simply are not structurally stable, or even qualitatively predictable,' says John Guckenheimer. The landscape model of Waddington and Thom had basins and valleys to 'attract' processes; Guckenheimer and others have identified new mathematical structures they call 'strange attractors' and 'vague attractors', which appear to lead neither to a stable state nor to a stable pathway of change. Only time will tell whether catastrophe theory can adequately describe the chaotic behaviour of systems caught by these attractors.

In addition to the questions of local v. global analysis and of structural stability, catastrophe theory faces a deeply rooted opposition from those who believe it is, as Zahler and Sussmann put it, an attempt 'to deduce the world by thought alone'.

These opponents see in Thom's philosophy a new version of the mathematical idealism of Pythagoras and Plato, which placed geometry on a higher level than our imperfect, changeable perceptions of nature. This charge hardly seems appropriate in view of Thom's explicit position : 'one should never state that, due to such and such a theorem, such and such a morphology is inevitably going to appear. In no case has mathematics any right to dictate anything to reality. The only thing one might say is that, due to such and such a theorem, *one has to expect* that the empirical morphology will take such and such a form. If reality does not obey the theorem – that may happen – [it] makes the situation all the more interesting.' Indeed, he slyly turns the tables on his opponents by pointing out that the most rigorous quantitative laws, those of gravitation and electromagnetism, are inextricably bound up with the geometry of space-time, as the theory of relativity shows. 'To require that all natural phenomena be ruled by a quantitative law is in fact to require that all phenomena be reducible to the geometry of space-time. I may be a professional geometer, but nonetheless I find this postulate a bit exorbitant.'

ARGUABLE ASSUMPTIONS

Two assumptions are needed to apply catastrophe theory as it now stands : first, that the system described be governed by a potential, and second, that its behaviour depend on a limited number of control factors. Without these assumptions, the classification of the elementary catastrophes is impossible.

As we have seen, the concept of potential is very widely used. The mathematical precision of the concept, however, varies from case to case. It is clear that a rubber band, for example, behaves so as to minimize tension, and there is general agreement on how to express that fact mathematically. But when Zeeman assumes in one of his public-opinion models that a democratic government acts so as to maximize public support for its policies, it is more questionable. 'You can always find or define something that looks like a potential,' says one mathematician, 'but not every potential has to give rise to an elementary catastrophe.' Mathematical biologist Jack Cowan contends that 'biology is full of oscillations and more or less stable cycles on all levels, and it's rare to find a straightforward potential that makes the situation strictly amenable to catastrophe theory'.

While these criticisms carry weight, it should be remembered that one form or another of the concept of potential has always been a part of science. In the third century A.D., Heron of Alexandria showed that light is always reflected so as to follow the shortest available path, and that idea runs throughout the history of optics. Other 'maximum' or 'minimum' principles are common in physics and chemistry. D'Arcy Thompson showed how frequently they play a part in shaping organic form. The most far-reaching theory in biology, that of evolution, is built around an implicit potential: the idea that evolutionary success is determined by an organism's ability to maximize its reproduction in a given environment. Economic theories focus on maxima and minima in relations of cost, production, utility and so on. So catastrophe theory's assumption of a potential in its applications, while arguable, is based in a long and successful tradition.

Similarly, the assumption of a limited number of control factors is part of all useful models, not peculiar to catastrophe theory. In some sense, of course, everything in the universe affects each particular event – but in practice we always restrict our consideration to a few important factors. Population geneticist Richard Lewontin puts it neatly: 'It may be true that "thou canst not stir a flower without troubling of a star", but the computer program for guiding a space capsule does not, in fact, have to take my gardening into account.' For all practical purposes, elementary catastrophe theory's limitation to processes with only a few control factors is hardly a restriction at all. Good judgement is required in selecting the factors for a specific model, though: if an important factor is omitted, catastrophe theory (like any other theory) will give a wrong or misleading picture.

NO HOLDS BARRED

The attraction of a new idea in science is eloquently expressed by Lewis Thomas in *The Lives of a Cell*. He likens scientists responding to a novel concept to insects that locate food or mates by homing in on minute amounts of airborne chemicals. 'As soon as a trace is released, receptors at the back of the neck are caused to tremble, there is a massive convergence of mobile minds flying upwind on a gradient of surprise, crowding around the source ... in a kind of activity that seems as random and agitated as that of bees in a disturbed part of the hive, there suddenly emerges, with the purity of a slow phrase of music, a single new piece of truth about nature.'

So far, there has been much more agitation than purity in the reactions provoked by catastrophe theory. Indeed, the controversy has taken on the aspect of a full-scale intellectual brawl, perhaps the largest one since the fight over the theory of evolution (another qualitative idea that was intended to organize and make sense of existing observations). Because catastrophe theory's range appears so great, sweeping claims have been made for it ... and those who do not share the enthusiasm have responded with equally sweeping denunciations.

Theories may be abstract, but mathematicians and scientists are flesh and blood. The lay public is becoming more aware of this: *The Double Helix*, for example, was a best-seller because it gave readers a behind-the-Nobel-Prize look at the fiercely competitive search for the structure of DNA by its author, James Watson, and his colleague, Francis Crick. Laboratory research and molecular model-building, though, are familiar by comparison with the blackboard explorations of mathematics. We can readily understand Watson's elation at learning that his rival Linus Pauling had made a mistake that would slow him down, but it's much more difficult to see how someone's reputation could be at stake in an esoteric argument over the validity of the Whitney C^∞ topology for certain function spaces.

The somewhat arbitrary division between pure and applied

mathematics also plays a part in the controversy. ' "Imaginary" universes are so much more beautiful than this stupidly constructed "real" one,' wrote pure mathematician G. H. Hardy, 'and most of the finest products of an applied mathematician's fancy must be rejected as soon as they have been created, for the brutal but sufficient reason that they do not fit the facts.' Keller, Sussmann and others have charged that catastrophe theory is attracting attention from pure mathematicians who want to do something useful 'without knowing anything about mathematics', and that the theory 'offers an opportunity for proving theorems, together with the gratifying feeling that what they are working in is "applied" '.

An applied mathematician is likely to become suspicious when he hears Zeeman's pleased description of the Bellaggio meetings : 'At the first one in 1966, each specialist talked to his nearest neighbour – pure mathematician to applied, applied to mathematical physicist, physicist to chemist, and so on. By the last one, pure mathematicians and biologists were talking directly to each other.' The applied mathematician is not accustomed to being leapfrogged this way, and his self-esteem may reinforce his opinions when he tells the pure mathematician, 'You don't know what the real world is like,' and tells the scientist interested in catastrophe theory, 'You don't know enough mathematics to see how flawed it is.'

Among both pure and applied mathematicians there is a degree of surprise and resentment at the widespread public interest in the theory. The *Science* article noted that catastrophe theory had been the subect of the first article on mathematics in *Newsweek* in at least seven years, referred to Zeeman as a 'publicist', and concluded, 'Although it is too late to nip the claims for catastrophe theory in the bud, Keller and others hope that the catastrophe theory craze will now begin to wane.' Sussmann and Zahler suggest that the theory's claim to universality, its formidable mathematics, and its 'impressive name' account for its success.

All the participants in the controversy are painfully aware that the finer points – including the meaning of the word 'catastrophe' in the theory – tend to be lost in non-technical accounts. Inevitably, the theory's name has focused public attention on the most spectacular discontinuities such as stock-market crashes. A front-page story in the New York *Times* was headlined EXPERTS DEBATE THE PREDICTION OF DISASTERS. Jonathan Rosenhead, a social scientist who attacked Zeeman's modelling of a prison riot as a step towards Big Brother, charged that 'catastrophists say they are simply studying instances of discontinuous behaviour ... [but] discontinuous change becomes labelled as catastrophe, and the status quo, by definition, becomes something to maintain'.

In fact, Thom's choice of a name for his theory was very subtly appropriate : it was inspired by the 'ultraviolet catastrophe' of classical physics. Pre-quantum theories of radiation predicted that electromagnetic waves in an enclosure should shift to higher and higher frequencies without limit, passing through visible light to the ultraviolet and beyond. This certainly would have produced a spectacular burst of some sort ... except that it didn't happen. That it didn't was a challenge to classical theories, and to explain the catastrophe that didn't happen, Max Planck was forced to postulate the existence of discontinuous quantum packets of energy instead of the smooth flow everyone had taken for granted. The multiple meanings – a non-existent theoretical catastrophe that provoked a conceptual jump in physics – must have pleased Thom when they occurred to him, but his supporters have come to regret the choice of name, if only for the misinterpretations it causes.

Misinterpretation is easier to correct than some of the misrepresentations and out-of-context distortions that have occurred. In a 1975 paper on the philosophy of science, Thom reminded his readers of Leibniz's dream of a truly rational, mathematical language, so well designed that to use it properly would be to reason without error. He suggested that his theory might be a first step in that direction, and speculated that if all concepts could be formulated mathematically, then all those who used them would be mathematicians. 'In such a vision of science, only the mathematician ... has the right to use the

mathematical concepts; only he has the right to be intelligent.'
Sussmann quoted this sentence in context; the *Science* article
repeated it without context; and a general-audience publication,
in a brief note based on the *Science* article, completed the job:
'... many non-mathematical scholars took up the theory in
spite of a modest assertion by its creator, René Thom, that in
the future "only mathematicians will have the right to be intelligent".'

Critics of catastrophe theory also charge that its expositors
deliberately intimidate non-mathematical audiences with references to 'deep' or 'profound' mathematics, although this would
seem a matter of personal opinion. Even among those who are
sceptical of the theory's applicability, many believe that
Thom's technical work merits those adjectives. Thom himself
has been chided for using topological terms in his writing for
linguists. It is hard to imagine what other terms the critics think
he should use, or what useful end would be served if he
abandoned the vocabulary of his life's work. As Alfred North
Whitehead, another mathematician-philosopher, said after a
talk on relativity, 'I regret that it has been necessary for me in
this lecture to administer such a large dose of four-dimensional
geometry. I do not apologize, for I am really not responsible for
the fact that nature in its most fundamental aspect is four-
dimensional.'

There are more legitimate objections to the way in which
the theory is presented. Zeeman's selection of processes to
model, for example, is clearly tendentious: he chooses those
which display bimodality, divergence and sudden transition,
and models them with the cusp. This may be ingenuous, but is
hardly surprising, since most of Zeeman's models are addressed
to readers unfamiliar with catastrophe theory. He wants to
show how it works, and naturally begins with cases in which
he already has reason to believe it does. In fairness, it should be
pointed out that just as Zeeman selects his examples, so do his
critics: they have concentrated on the most speculative models,
ignoring the most rigorous ones which are harder to fault.

Probably no criticism of the way the theory has been presented is more pointed or well-informed than John Gucken-heimer's. Thom and Zeeman, he says, show 'a real reluctance to get their hands dirty with the scientific details of the applications'. Thom is interested above all in extending the theory towards his original goal of modelling the structural stability of life, and has had little contact with experimenters who would like to put his ideas to the test with single cells or embryos. At times, his dissatisfaction with established theories has led him to ill-advised sarcasm, as when he called contemporary biology 'a vast graveyard of facts'. It was from facts, after all, that Waddington and others drew the ideas which stimulated Thom in the first place.

Zeeman, although he has proposed so many models, is more concerned with stimulating others to test the theory than with doing so himself. A painstaking cycle of trials and modifications may be needed to fit a single model to the unyielding facts. In response to criticism of his nerve-impulse model, Zeeman has suggested modifications that would improve it, but so far has not published them. Asked early in 1976 if he was satisfied with the progress of applied catastrophe theory, he replied, 'If anything, it's all going too fast. I'm interested in more than I have time to keep up with.' Time may prove him right in believing that as a mathematician he can do more for the theory by making it accessible to scientists than by applying it himself. Until then, Guckenheimer's criticism carries weight.

In the correspondence that followed the article in *Science*, the controversy began to turn in on itself : the issue became one of the catastrophe theory 'craze' and the backlash, rather than the theory's merits. Biologist Robert Rosen counselled moderation : 'While it must be granted that a number of immoderate claims in the form of "catastrophe theory can do everything" have been made in the literature, on the basis of too little experience, it doesn't seem that the proper response is an equally immoderate claim that "catastrophe theory can do nothing" on the basis of that same body of experience.' Marc Lewis, a psychologist who is developing statistical techniques for developing catastrophe surfaces to fit experimental data, wrote : 'Newton's critics eventually caused him to call science a series of lawsuits. Now Thom's admirers give us a modern Newton and his detractors give us new lawsuits. Let us have neither.'

Sussmann replied, 'Do they believe in the general postulate that, given any disagreement, the truth must lie in the middle? Such a claim seems clearly fallacious. If you believe in Nazism and I believe in democracy, how many will argue that the truth must lie halfway between us?'

His choice of an example raises another question : is catastrophe theory a threatening ideology? Clearly not, unless one accepts the ludicrous portrayals of Zeeman as a prison tyrant or of Thom as a pretender to exclusive intelligence. But on a deeper level, the theory may indeed be threatening because of the proliferation and variety of its models. One of the critics, mathematician Mark Kac, wrote an intriguing article on the role of mathematical models in science nearly ten years ago. In it, he said that 'the main role of models is not so much to explain and to predict – though ultimately these are the main functions of science – as to polarize thinking and pose sharp questions. Above all, they are fun to invent and play with, and they have a peculiar life of their own. The "survival of the fittest" applies to models even more than it does to living creatures. They should not, however, be allowed to multiply indiscriminately.'

And so we come full circle, to evolution. Certainly the models of catastrophe theory have polarized thinking and posed sharp questions, although perhaps not in the way Kac had in mind. And some people, at least, have found that these models have a life of their own. But the survival of the fittest, in fact, *does* mean indiscriminate multiplication; after all, it is the competition among models to survive and to reproduce – to gain a lodging in science and give birth to new and better models – that leads to natural selection. Kac seems to suggest that some form of artificial selection is necessary – but on what grounds? Because unsatisfactory models are harmful, or because good models cannot be trusted to displace unsatisfactory ones as scientists learn what is useful and what is not?

One can only wonder what Sussmann and Zahler mean when they write, 'It is Zeeman's theory that poses the most immediate danger', or who is being protected when they write, 'We feel that the many researchers now being attracted to catastrophe theory stand to gain nothing but wasted time.'

Our own position is that suggested by Robert Rosen: 'If an individual scientist finds such concepts uncongenial, let him not use them. There is no reason why he should take their existence as a personal affront.' It seems likely that catastrophe theory will survive the current controversy, at least if its most vehement critics will concede that science as an enterprise can survive catastrophe theory. In the long run, the theory's impact will depend not on the arguments for and against it, but on how useful it proves to be.

A Note on the Applications

THE chapters that follow present applications of catastrophe theory to many fields, in the form of models based on the elementary catastrophes (usually the cusp and the butterfly). The applications range from natural science to politics, from relatively 'solid' to frankly speculative. Some are Zeeman's, a few are due to others, and most are original. The criticism that has been levelled at some of them will be taken up where appropriate, and their possible advantages outlined.

In some cases, the examples may seem obvious or trivial. Keep in mind that they are by no means representative of the full scope of catastrophe theory; they exploit only those properties of the elementary catastrophes which can be made clear in a non-technical introduction like this book, and their function is to stimulate your thought, not to break new ground in either catastrophe theory itself or science.

These models are *descriptions* – descriptions that frequently help to make more sense of the processes they describe – rather than explanations. If they suggest qualitative predictions, well and good; but they are meant to be judged not as guides to new discoveries, but as examples of a way to 'turn old fact into new knowledge'. They are rough sketches from an art of models.

5

APPLICATIONS IN PHYSICS, CHEMISTRY AND BIOLOGY

> *'If you want to know what happens when you throw a stone into a pond, it is infinitely better to make a trial and film it than to attempt to theorize about it; the best specialists [in fluid dynamics] would certainly be unable to tell you more about it'* – RENÉ THOM

IN thinking about science, we traditionally draw a line between two groups of disciplines : the exact, 'hard' sciences of natural phenomena and the inexact, 'soft' sciences of social and cultural phenomena. To most people, the first group seems more certain, more down-to-earth, more realistic, because it can offer laws of nature while the second group has only generalizations. Even some social scientists tacitly share the belief that with time, with better data, and with better mathematical models, their relatively young disciplines will also become like natural science – exact and predictive.

As any honest natural scientist will admit, however, the laws of nature derive their power, simplicity and elegance from their *lack* of down-to-earth realism. Galileo and Newton, for example, were able to analyse motion so effectively because they postulated that the 'normal' type of motion was straight-line, unchanging, eternal (unless disturbed) motion – something which has never been observed and never will be. That kind of fruitful idealization is at the heart of mathematical science. As C. P. A. Pantin pointed out in 1968, 'Physics and chemistry have been able to become exact and mature just because so much of the wealth of natural phenomena is excluded from their study.'

Thus the natural sciences that are most highly developed mathematically are those in which idealization has gone farthest. In quantum physics, for example, a formula called the 'Schrödinger equation' expresses everything there is to know about the energy states (and therefore the possible behaviour) of an atom. Then why, as Thom implies, do we understand almost nothing of the intricate forms created when a stone splashes in a pond? Because the Schrödinger equation yields explicit answers only for the simplest atoms. In principle, we could use a vast array of these equations to describe the reaction between hydrogen and oxygen atoms that produces water, the physical properties of water, even the splash. In practice, however, that description would tell us nothing useful, since we cannot *solve* the array of equations. Waddington once said wryly that from a biologist's – or even a biochemist's – point of view, the equations are 'a trifle on the inscrutable side ... if they do contain everything necessary to understand protein molecules, no one can get it out of them'.

Nowhere is the gap between mathematical theory and even the most common phenomena more striking than in Thom's area of special interest: morphogenesis, the origin of form in both life and inorganic nature. It was to help close that gap, above all, that Thom created catastrophe theory. And although most of the widespread interest in catastrophe theory has centred on its possible applications in the social sciences, the theory has already made solid contributions in the physical sciences. There it can be every bit as rigorous as more traditional theories, because the governing potential and the control factors can be clearly defined. In optics, for example, the curves of the elementary catastrophe surfaces can even be seen and photographed. Under another name, that of caustics, they have been known and studied for centuries.

CAUSTICS AND CATASTROPHIES

Imagine a group of light rays, all travelling in straight lines. The rays leave a point source of light and encounter something which reflects or refracts (bends) them – a mirror, a lens, a water droplet, a layer of hotter or colder air. In the ideal (i.e., mathematically most manageable) case, all the rays are re-directed to another common point, the focal point. The result is a sharp image, and so this ideal case underlies the whole of practical optics : the design of cameras, telescopes, microscopes and so on.

In nature, though, the ideal case is almost never realized. Instead of causing the light rays to converge at a point, the reflection or refraction is far more likely to direct them so that they converge (if at all) on a line, or fan across a surface, or even form something as beautifully complex as a rainbow. An early observer of these complex patterns noticed that the light was often concentrated along bright lines which he named 'caustics' for their burning intensity. The focal point is a special case of a caustic; it is a cross-section of a conical 'envelope' of rays, so that if a projection screen is moved towards or away from the lens that is projecting a point of light onto it, the point broadens into a disc. To classify and explain the far more complex shapes of all possible *non*-ideal caustics and envelopes has taxed the abilities of some of the greatest scientists and mathematicians, among them Gaston Darboux, whose work in geometrical optics influenced Thom.

As Thom realized very early, caustics are natural subjects for catastrophe theory. The path of light rays is governed by a potential (rays always follow the shortest path between two points). The bright and dark regions of a caustic projected on a screen are determined by maximum and minimum values of the formulae of optics. And the intensity of the light predicted by optical theory reaches a maximum along a certain line and drops discontinuously to nothing beyond that line. In effect, then, a caustic is a spatial arrangement of maxima and minima for a system governed by a potential – and that is precisely what the catastrophe graphs are.

In laboratory experiments with caustics, physicist Michael V. Berry of Bristol University has shown that catastrophe theory accurately characterizes and predicts the shapes that appear in many natural and artificial caustics, and that it can add to our understanding of phenomena as diverse as the twinkling of starlight and the scattering of particle beams by crystal surfaces. Working with J. F. Nye, Berry investigated the caustic junctions formed where several bright lines of light come together, as they do at many points in the rippling network of brightness on the bottom of a swimming pool. On casual examination, many of the junctions appear to be three-way meetings – an observation that seems not to have surprised anyone previously, since as Berry and Nye point out, triple junctions are very common in nature, 'in mud cracks, foams and the markings on giraffes', for example. But from the perspective of catastrophe theory, triple junctions in this case *are* surprising – in fact, they are impossible, since no cross-section of the applicable catastrophe surfaces produces such line patterns. So Berry and Nye set up apparatus to produce and photograph caustic junctions in the laboratory. Through a microscope, they observed previously unsuspected fine structure in the junctions : what had appeared to be triple junctions were in fact curved triangles with exactly the topology expected on the basis of catastrophe theory.

Figure 15a is a computer drawing of a caustic produced by a laser beam passing through rippled glass like that often found in bathroom windows. Its four cusped corners, partially enclosed by curves, are four sections of the hyperbolic umbilic catastrophe surface. The catastrophe itself is the discontinuity in brightness at the boundaries of the caustic. It would appear as the 'jump' of the needle on a lightmeter as the meter's sensor crosses the boundary. The caustics from other panes of rippled glass would be all of the same topological type, because the cusped corners and curves are structurally stable, although their quantitative features would differ. In a special case, where the refracting ripples of the glass are nearly symmetric, the caustic becomes a nearly square form like the computer drawing in Figure 15b. Special cases such as this are non-generic – that is, any disturbance of the symmetry causes them to change their topological type – and thus, while they play a large role in traditional geometrical optics, they are not structurally stable and are comparatively rare in nature.

So the apparently abstract scheme of catastrophe theory is

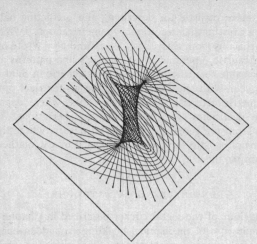

Figure 15a.

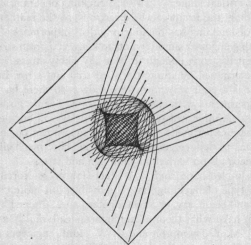

Figure 15b.
Projections of a catastrophe surface

actually more realistic for describing and predicting naturally occurring caustics than traditional geometrical optics. 'We draw two conclusions from this study,' Berry and Nye wrote of their work on caustic junctions. 'First, nature's line patterns are not all of the same sort; the triple junctions generic in mud cracks cannot occur with caustics. Second, the geometrical optics of ... symmetric artifacts such as telescopes, where departures from the ideal point focus are treated as "aberrations", is very different from the geometrical optics of nature, where the generic forms of caustic surfaces are governed by the mathematics of catastrophe theory.'

ENGINEERING CATASTROPHE

The behaviour of the metal clicker described in Chapter 3 is a very simple case of mechanical buckling : a sudden change in the shape of a more or less elastic structure when the stress on it reaches a critical value. When such buckling occurs in a bridge or a building, the result can be disastrous, so the mathematical analysis of buckling has obvious practical importance. Usually the working engineer simply wants to know if a given structural component is strong enough to bear any likely stress, and that can be determined by straightforward trial and error. But trial and error is a poor way to design bridges, so there have been many experiments with carefully fabricated beams, girders and panels to determine just when and how they will buckle under stress. These experiments form part of theoretical engineering, which aims to deduce general laws about how mechanical behaviour depends on variations in design and stress.

For a simple shape such as a steel slab or reinforcing rod, only one type of buckling may be possible. But more complex shapes – an I-beam, say, or a rectangular panel stiffened by cross ribs – can have what theoretical engineers call 'multiple failure modes'. J. M. T. Thompson and G. W. Hunt, engineers at University College in London, applied bifurcation theory to the study of this type of problem in their 1973 book, *A General*

Theory of Elastic Stability. They treated each possible type of buckling as one of the solutions for a stress equation whose graph branched into two or more distinct curves. Subsequently they became aware of catastrophe theory, and in an article in *Nature* in 1975 they wrote: 'The two theories have been developed quite independently, but are remarkably similar in form and content. Cross-fertilization between the two schemes promises to be most rewarding.'

In one experiment described by Thompson and Hunt, their colleague J. Roorda tested the strength (resistance to buckling) of a simple model arch subjected to loads at various points on its upper surface. His results (Figure 16a) show that strength falls off very sharply when the load is applied even a small distance from the arch's centre: a displacement of only 1 per cent of the total length of the arch causes a 10 per cent reduction in strength. The same sensitivity can be expected of an arch which is loaded at the centre but has some degree of asymmetry, due to unavoidable imperfections of the manufacturing process. The sharply peaked curve in Figure 16a is known as a cusp; it can be derived from the cusp catastrophe surface by projecting the fold lines in that surface (the 'lips' of the pleat) on to a horizontal plane. As we have seen, each of those points is the brink of a catastrophic jump in the system, just as is each of the experimentally determined points on the graph here. Using calculus, one could determine the slope of this line at every point *except* the singularity, the 'peak' at which it suddenly changes direction. It is this kind of mathematical situation which so often stymies Newtonian mechanics, and for which bifurcation theory and catastrophe theory have been developed.

Figure 16b reveals that the higher-dimensional catastrophes also can occur in more complex elastic buckling. It shows a three-dimensional graph of the strength of a stiffened panel in terms of two different manufacturing imperfections. Hunt plotted the graph on the basis of bifurcation theory; however, he wrote later with Thompson, 'it has been identified by David Chillingworth of Southampton University as a hyperbolic–

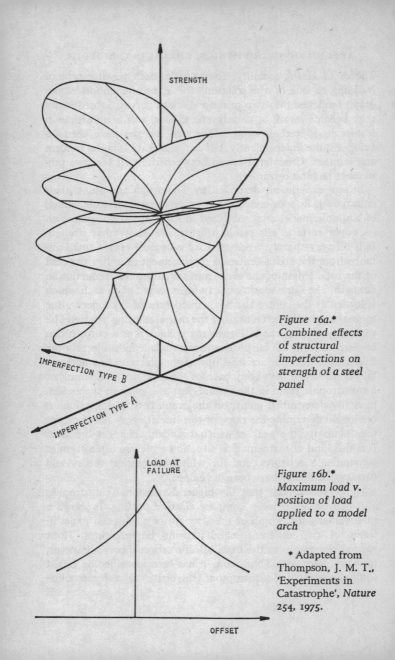

STRENGTH

IMPERFECTION TYPE B

IMPERFECTION TYPE A

Figure 16a.*
Combined effects
of structural
imperfections on
strength of a steel
panel

LOAD AT
FAILURE

OFFSET

Figure 16b.*
Maximum load v.
position of load
applied to a model
arch

* Adapted from
Thompson, J. M. T.,
'Experiments in
Catastrophe', *Nature*
254, 1975.

umbilic catastrophe, and this seems to be the first known practical example of this predicted form'. Its curves reflect the likelihood of several divergent types of buckling.

Thompson and Hunt note that many design engineers are currently using a technique called 'simultaneous mode design', in which the optimal structure (the lightest or most efficient component to perform a given task) is that which theoretically would fail in all modes at once. In other words, it has no excess of strength in any direction; rather like the proverbial 'one-hoss shay', when it gives way it does so completely. This is sound, economical design – but as Thompson and Hunt caution, it also produces structures that are exquisitely sensitive to imperfection. It is not enough to know the critical stress, that is, the quantitative breaking point of a complex design; one should also know as much as possible of the qualitative geometry of its failure modes, because what will happen beyond the critical stress level can be very different from one case to the next, depending on just which path the buckling takes. And here catastrophe theory, joined with bifurcation theory, can be very helpful by indicating how new failure modes appear. The theoretical engineer can use the mathematics of both these theories to 'home in' on regions where behaviour is likely to be discontinuous rather than smooth, instead of simply hoping that all possible failure modes have been discovered in tests of a dozen or a hundred or a thousand samples. The mathematics may be less familiar than the idealized analysis of buckling descended from Euler, but then engineers must work with real structures that are neither ideally symmetrical nor ideally uniform in strength, and it is certainly better to adopt new mathematics than to discover a new, non-ideal failure mode in a completed building.

PHASE TRANSITIONS

One of the most successful early generalizations of modern physics was advanced by the seventeenth-century Irish scientist Robert Boyle, who proposed on the basis of experiments that the pressure and volume of a gas sample at constant temperature are inversely proportional – that is, twice the pressure produces a compression to one-half the volume, and so on. Boyle's Law was soon extended to cover changing temperature as well; higher temperature brings an increase in volume, pressure, or both, and lower temperature a decrease. Figure 17a shows a three-dimensional graph of the three quantities. The isotherms (solid lines) trace changes in pressure and volume at several fixed temperatures; the isobars (dashed lines) trace volume and temperature at fixed pressures; and the isometrics (dotted lines) show pressure and temperature for three fixed volumes.

Unfortunately, scientists later realized that Boyle's Law is true only as an idealization, and that their experiments would never yield the same smooth lines. Figure 17a illustrates the behaviour of an *ideal* gas, and no real gas – let alone the mixtures of gases in our lungs, in auto engines or in a thundercloud – behaves exactly that way. Worse yet, high enough pressures and low enough temperatures can bring about a phase transition, converting any gas to a liquid or solid. At the transition points the isotherms, isobars and isometrics are not well-behaved at all, and so Boyle's Law has given way to increasingly complex equations. A scientist working with a particular substance must 'plug in' experimentally determined correction factors to make these equations fit. Figure 17b, for example, shows the phases of water observed in high-pressure experiments : there are seven different types of ice, each with its own molecular structure and properties, some found alone and some coexisting with liquid or vapour. Of course, it is possible to develop an equation that fits this surface more or less accurately – but such a long, unwieldy equation isn't much use to the climatologist working on iceberg formation (water plus salt plus dissolved oxygen) or the biologist who wants to know how microbes can survive freezing (water plus dozens of other substances).

At the last Bellaggio conference, David Fowler of Warwick showed how Thom's mathematics could be connected to the equations often used by scientists studying phase transitions. The model of boiling and condensation in Chapter 3 (Figure 9) is a highly simplified example of that application. Its simplicity

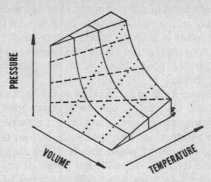

Figure 17a. A three-dimensional graph of the behaviour of an 'ideal gas'

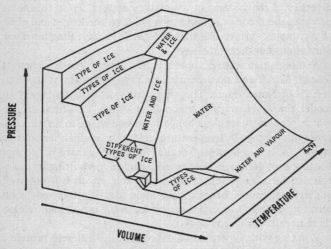

Figure 17b.* A three-dimensional graph of the behaviour of a real substance: water

* Modified and redrawn from *Physical Chemistry*, 3rd edition, by Walter J. Moore, Prentice-Hall, 1962.

renders it inaccurate in many respects, but not necessarily less useful for that reason : as Zeeman says of Boyle's Law, even a generalization that is patently false in many cases can be valuable if it contains insight into varied phenomena and provides an 'intellectual resting point' from which we can proceed with later modifications.

L. S. Schulman, a physicist working at the Technion in Israel and at Indiana University, has used catastrophe theory to describe transitions in magnetic and other properties caused by changing temperature. He believes the theory can be integrated with existing theories of phase transition for many materials, and that while it sacrifices predictive power, it offers compensating advantages by classifying different types of transitions and their stability. 'It might be said', he remarks, 'that the real problem of phase transitions is a qualitative one, namely, to explain the mutual resemblance of phase transitions despite the diversity of the systems in which they appear.' From this point of view, the theory can serve as a guide to the ledges and cliffs of a complex surface like that in Figure 17b, revealing not their dimensions but their topology.

A more quantitative approach has been taken by chemist John J. Kozak and mathematician Craig J. Benham at the University of Notre Dame. They have used catastrophe theory to model the biochemical phenomenon called 'denaturation', in which protein molecules change from one shape to another because of a change in temperature, acidity or some other control factor. (A few minutes of boiling produces a hard-boiled egg because heat denatures the protein albumin, permitting its molecules to form new links. A more subtle control, still not well understood, causes haemoglobin molecules in your red blood cells to clamp on to oxygen molecules in the lungs, then release them to cells within the body.)

Working with experimental data on several different proteins, Kozak and Benham constructed 'model' potentials for denaturation – equations which, while not necessarily producing the same curves as the actual (and unknown) governing potentials, had minimum values at appropriate levels of the control factors. Using a statistical approach that reflects the uncertainty of denaturation for any given molecule among trillions in a sample, they were able to make quantitative predictions that agree well with observation. In one case, they noted a sign of possible hysteresis in the transition from one molecular form to another : a protein found in *staphylococcus* bacteria requires slightly less acid conditions for denaturation than for the reverse transition. Catastrophe theory, they concluded, may make it possible to draw 'phase diagrams' analogous to Figure 17b for biochemical transitions, and to determine theoretically relationships impossible to measure directly – in a living cell, for example, where it might be impossible to vary the temperature without disrupting the process entirely.

The work of Kozak and Benham drew sharp criticism from Zahler and Sussmann in their *Nature* article in late 1977. The critics contended that Kozak and Benham were confusing a sudden but continuous change in shape with a mathematical discontinuity; that a number of other behaviour surfaces (including one that they suggested) would fit the data just as well; and that it is absurd to believe that one can deduce a system's behaviour under all circumstances from a few laboratory measurements.

As for the first objection, Kozak and Benham's statistical approach explicitly recognizes that a change in form for even one molecule, let alone trillions, cannot be absolutely discontinuous. The point, though, is that denaturation is a qualitatively sharp and well-defined event in an otherwise smooth process. John Guckenheimer, who joined Zeeman, Berry, Brian Goodwin and others in replying to what he called the critics' 'snide, mean-spirited' attack, points out that the linking of mathematical discontinuity and observed 'jumps' in natural processes is not limited to catastrophe theory, but is 'a common and useful approximation in many mathematical models'.

The choice of the cusp catastrophe behaviour surface for the Kozak–Benham model, which Zahler and Sussmann profess to find mysterious ('So we must ask again : what is special about a model that looks like a cusp?'), is dictated by the fact that the cusp catastrophe is structurally stable. Zeeman's reply is short and to the point : 'The simple answer to their question is that the cusp catastrophe is stable, whereas [Zahler and Sussmann's] figure is not.' There is no mathematical reason to believe that the proposed alternative surface would remain qualitatively stable if perturbed, and since denaturation in fact does show qualitative stability in a wide range of circumstances, the preference for a stable model is only logical.

As for the third criticism – the impossibility of deducing global behaviour from local behaviour – it is hard to understand in view of Kozak and Benham's statement : 'The model presented is local in that it describes the behaviour of a system [in] some range of temperature, pH [acidity] or other controlled variables. In other ranges different transitions may occur. A global picture of the overall behaviour of a given protein-denaturant system requires linking together each (local) model of each transition undergone by the system.' In other words, they propose to extrapolate behaviour from observed transitions *only* as far as experimental data allow, leaving room for a new or modified scheme if new types of denaturation are observed. Zahler and Sussmann, by contrast, appear to be saying that since other types of transition may exist, it is wrong to extrapolate at all – an attitude that may be logically impeccable, but has little to do with science. The critics deride the Kozak–Benham model as 'too good to be true'; indeed it is, as is every general law of natural science.

BIOLOGY

Catastrophe theory is being applied to biology at many levels, from biochemistry to genetics, embryology and evolutionary theory. At each level it may help to bridge the gap between what we know of life's detailed microscopic workings (usually all too little) and what we can see of its overall qualitative organization (usually much more than could be expressed mathematically, at least until now). In his foreword to Thom's book, Waddington described this gap clearly :

When a category of biological processes, such as evolution or development, leads to the formation of [a] specifically biological body of theory [rather than physical or chemical explanations], it does so because it exhibits two characteristics; it involves entities which have a certain global simplicity and definiteness of character (e.g., a given species of animal or plant, an organ such as the heart or liver, or a cell type such as a muscle or nerve cell), but if one attempts to analyse these entities into basic constituents, such as genes or molecules, they turn out to be of unmanageable complexity ... If there was no simplicity, there would be nothing to make a theory about; if the complexity remained manageable, physico-chemical theories would suffice ... Thom has tried to show, in detail and with precision, just how the global regularities with which biology deals can be envisaged as structures within a many-dimensional space.

Under certain circumstances, Thom believes, the shapes of the catastrophe surfaces may appear in biological forms just as they do in caustics. He draws attention to suggestive similarities between the unfolding of the hyperbolic umbilic catastrophe, for example, and the formation of a mushroom's stem and cap. Figure 18 shows two visually similar 'infoldings' : those of a sea-urchin embryo and of the elliptic umbilic catastrophe. Is there a fundamental correspondence between these forms, or is the similarity coincidental? Most biologists are suspending judgement, for it will take years of experiment to develop detailed models correlating the control dimensions of the catastrophes with the three dimensions of space and one of time in which living morphogenesis takes place.

What is most exciting about the prospect of these models is that they are independent of time scale. If they can indeed capture the stability of complex changes of form, they can do so with equal ease for changes taking a few seconds or minutes within a cell, for changes taking a few days or months in an embryo, or for changes taking millions of years in evolution. Consider, as Thom does, the marvellous coadaptation of flowers

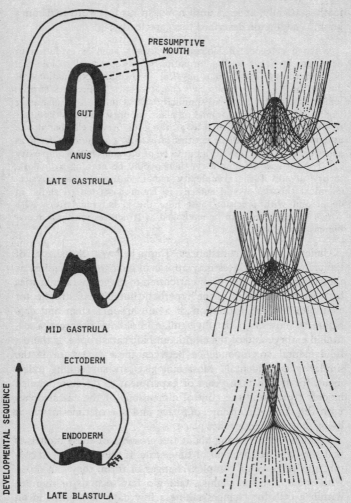

Figure 18. Two 'infoldings' – a sea urchin embryo and the elliptic umbilic catastrophe

and of the insects that feed on them. It is inescapable that they literally fit together, sometimes so precisely that a plant can be pollinated by only one species of insect among thousands. Must we wait until we know the genetic code – and the far more challenging and obscure developmental code – for both plant and insect, or can we speculate with Thom that whatever the potentials involved, whatever the myriad control factors, there is a topological 'image' of the insect in the flower's genes and vice versa?

A pessimistic point of view holds that even the simplest biological form involves so many different factors that the elementary catastrophes (limited to four control factors) cannot possibly be applied. But over and over again, biologists have found that very complex processes can be started or stopped by simple stimuli (as Waddington and his colleagues found while studying chemical influences in embryology). All of the changes that transform an insect larva into its adult form within the cocoon, for example, are triggered by the presence of a single hormone. And although the nervous system of an animal or human being can convey an infinite variety of messages, the transmission of every nerve impulse depends on appropriate levels of just a few substances (especially sodium and potassium) inside and outside each nerve cell. Key features such as these can serve as control factors in models based on catastrophe theory, permitting description – and perhaps even prediction – of qualitative changes far too complex for modelling by traditional methods.

As Thom points out, we can see much more than we can say: a film of a stone splashing into a pond, or even a still photograph of the splash, exhibits a pattern that cannot be thoroughly explained by the most powerful mathematical techniques of fluid dynamics. We have an elegant theory of crystallization – indeed, we have a number of them – but we don't really know what shapes the feathery branching of frost crystals on a cold window. Catastrophe theory opens up new ways to approach natural phenomena such as these. It owes much to the visual sensitivity of Thom and his predecessors in the study of morpho-genesis, and as time goes by it may well repay that debt by giving us better ways to say what it is that we have been seeing all along.

6

APPLICATIONS IN ANIMAL
BEHAVIOUR

*'The topology of this universal model [for biology]
will reflect less the [evolutionary] relationships than
the functional interaction between species, so that
the distance between bee and snapdragon will be
less than that between bee and butterfly'* – RENÉ
THOM

IF we could draw a map of the sciences, ethology – the study of
behaviour – would occupy a region bordering biology on one
side and psychology on the other. As long ago as 1872, Darwin
considered how evolution might shape behaviour. (His book on
that subject, *The Expression of the Emotions in Man and Animals*, is still stimulating reading.) But many of his ideas lay
fallow until the twentieth century, when ethologists such as
Niko Tinbergen and Konrad Lorenz renewed the subject with
superb field studies of animal behaviour. Popular works by
Lorenz (*King Solomon's Ring*, *On Aggression*), Robert Ardrey
(*African Genesis*) and Desmond Morris (*The Naked Ape*) have
attracted a wide audience. In Darwin's time it was controversial to assert that human beings and chimpanzees had common ancestors; today scientists and laymen alike take it for
granted that what we learn about aggression, territoriality and
family structure among primates and other animals can tell us
much about ourselves.

Many kinds of theories of animal behaviour have been developed. One kind is essentially parallel to theories of human
psychology : either Freud's, which models behaviour in terms of
innate drives and instincts, or the behavioural psychology of
Pavlov, Watson and Skinner, which treats behaviour as a complex of stimulus-and-response patterns. Both have a long and

influential tradition, but both are considered by many ethologists to be inflexible and restrictive – the first because it attempts to answer so many questions in terms of what cannot be seen; the second because it attempts to answer the same questions as if animals and human beings were the 'black boxes' of electrical circuit design, devices whose pattern of 'input' and 'output' is their only relevant feature.

Another general approach to understanding behaviour is that of traditional physiology and neurophysiology, which seeks the origins of behaviour in metabolic states and organic responses to chemical and neural signals. This approach has led to the identification of the brain's 'sleep centre', 'pleasure centre' and the like; to the isolation of hormones and neurotransmitters with far-reaching effects; and to scores of detailed models for biochemical cycles and neural networks underlying specific aspects of behaviour. The successes of this approach are numerous, but so far it yields a complete and coherent picture of behaviour only for the simplest organisms.

A third approach, that now known as sociobiology, is the more or less direct descendant of Darwin's thought. It takes for granted, as Konrad Lorenz puts it, that 'behaviour patterns are just as conservatively and reliably characters of species as are the forms of bones, teeth or any other bodily structures'. Its central questions are evolutionary : how did the transition from life in the trees to life on the lush African plains affect the behaviour of our hominid ancestors? How could self-sacrificing, altruistic behaviour confer a selective advantage on a species, even if it meant death for some individuals whose genes would thus be lost? In sociobiology, the 'Why?' of behaviour becomes 'How did behaviour evolve through adaptation?'

Each of these general approaches has its quantitative and predictive aspects. The stimulus-response theories have produced 'learning curves', showing how animals' performance on a variety of tests improves with repeated trials. The physiological theories lead to experiments in which surgical or chemical changes are tested for measurable effects on behaviour. Sociobiologists have demonstrated with the mathematics of heredity that in many cases an animal that sacrifices itself to protect its kin 'survives' through the genes it shares with them. But the variety and complexity of animal behaviour are so great that an adequate *description* for many phenomena still eludes us. It is here that catastrophe theory can be useful, because it enriches our conceptual vocabulary of cause and effect, and allows us to organize information in a qualitatively precise, multi-dimensional way. Catastrophe theory cannot provide new facts about behaviour, but it can help us to visualize the interaction of two or more factors shaping behaviour.

In this chapter, examples of territorial behaviour, aggression and group formation will be modelled using the cusp catastrophe graph to illustrate observed patterns of behaviour and their changes from one mode to another.

TERRITORIALITY AND PENDULUM FIGHTING

Territorial behaviour shapes the geographical distribution of different species and of individual members of single species. One of the broadest rules of ecology and population biology, Gause's Principle, asserts that no two species can coexist for long in a given region if they make precisely the same demands on the environment. A 'territory' can be very subtly delimited: field studies have shown five species of mosquitoes, for example, seeking their food and mates at different heights above the ground in a tropical rain forest. Within a single species, territorial behaviour defends an animal against others who try to encroach on a valuable resource such as a hunting ground or a nesting site. Robert Ardrey, in *The Territorial Imperative*, suggests that inherited patterns of territorial behaviour underlie human patriotism and warfare.

The patterns of territoriality seem to depend on both the resource for which animals are competing and the cost of defending it, in terms of the energy required to deter or fight off competitors. Often, actual fighting is rare because territorial boundaries are staked out by display behaviour. Many songbirds, for example, stake out their territories by repeating certain patterns of notes in trees at the boundaries of their turf. An interesting example is the 'lek' behaviour exhibited by grouse. These birds spend most of the year in loose social groups, often predominantly of one sex. As the breeding season approaches, the male grouse begin their display – stylized strutting and ruffling of their feathers. This establishes an area of dominance, a 'lek', to which the females come for mating. Afterwards, the females build nests and rear the young, generally at some distance from the lek. In this case the competition for other necessities, such as food, is subordinate to the competition for prime breeding areas. The same is true for Brandt's cormorant, a cliff-dwelling seabird that feeds in the open ocean. There appear to be no well-defined fishing territories, and there is no competition among Brandt's cormorants as they seek schools of fish. But their nesting sites on cliffs at the shore are often in short supply, and competition for them is intense.

Figure 19 shows territorial behaviour as a cusp catastrophe surface, with the level of competition (based on population density) and the cost of defending a territory as control factors. This model suggests that a low level of competition in itself does not encourage territoriality when the cost is high (point *a*). If a small population of Brandt's cormorants had access to extensive fishing grounds, there would be little incentive for any bird to claim a region as its own, because it would have to spend all its time patrolling the borders. If both competition and its cost are high (point *b*), territorial behaviour is more likely, but not fully developed: it may take the form of a hierarchical organization, a 'pecking order' such as that of barnyard fowl, in which the higher-ranking birds can drive off their inferiors from any location, but no permanent territories are claimed.

If the cost of defending a territory decreases for a species with a hierarchical structure, but competition remains high, territorial behaviour will develop further in a smooth (noncatastrophic) transition (path *b–c*). But if a non-territorial species, adapted to conditions of low competition and low cost of defence, is subjected to increasing competition (say, by a population

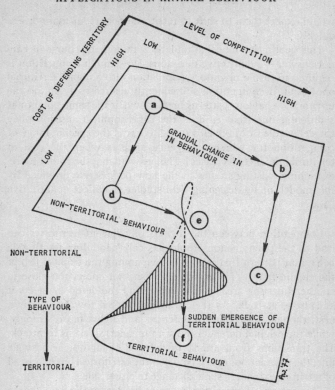

Figure 19. Territorial behaviour

explosion or by a climatic change that causes a decline in a necessary resource), then the model predicts a sudden (catastrophic) transition to territorial behaviour (path *d–e–f*). One would expect, then, that if a chicken breeder stepped in to remove every challenger to the dominant roosters, their descendants would show gradually increasing attachment to specific locations. But if, say, migratory birds which normally had a wide choice of nesting sites were confined to a zoo's aviary, one would expect them to shift to territorial behaviour more or less quickly.

This qualitative model stimulates – though, of course, it cannot answer – quantitative questions. How much competition is required to turn a migratory population into a highly territorial one? And, if competition subsequently decreases again, do the territorial behaviour patterns break down at the same point or at a different one (that is, does the catastrophic transition obey the delay rule or Maxwell's rule)? What is the specific range of competition for which divergence appears when the cost of territorial defence decreases? The answers may be found by re-examining existing studies in the new perspective provided by this model, or by designing field studies to collect quantitative data.

Competition between the owners of adjacent territories can lead to the phenomenon ethologists call 'pendulum fighting', a behaviour pattern that is well-known among the tropical fish of the cichlid family, for example. A fish that comes too close to another's nesting area is an invader, and the defender meets it and drives it back. As the invader is forced towards its own nest, though, it becomes a defender and fights more strongly, while the original defender (now an invader) slackens its attack. The transitions from attack to defence occur not at a single line, the abstract boundary between the territories, but in a band of territory extending some distance to either side.

This pattern of transition suggests the hysteresis loop shown in Figure 20. Here the control factors are the intensity of the territorial threat (the closeness of an invader to the nest) and the level of defensive motivation (varying with the physical condition of the animals involved, the phase of the nesting season, etc.). When both the invader and the defender are highly motivated, the hysteresis loop (f–g–h–i) is large. The model suggests that as the combatants tire pendulum fighting should 'fade away', the transition points becoming closer and eventually joining as the fight breaks off. This changing pattern has, in fact, been observed in cichlid fish and other species. (This model,

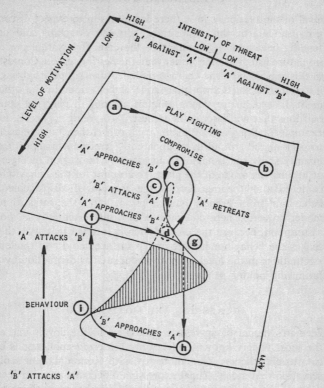

Figure 20. Territorial conflict: pendulum fighting

dealing as it does with individual territorial behaviour rather than that of a whole population of animals, justifies the use of the delay rule, which makes hysteresis possible.)

At an intermediate level of motivation, the alternating behaviour need not lead to a hysteresis cycle. Path *e–c–d–e* depicts a single encounter in which neither participant is strongly threatened. Here, the fish may simply keep an eye on each other, with perhaps an occasional flurry of fighting. And if the level of motivation is low, there is no reason to expect discontinuous (catastrophic) transitions between attacking and defending behaviour. This might be the case, for example, with immature fish or at times other than the nesting season. Contests in this region show the characteristic features of 'play fighting', with smooth shifts from one mode of behaviour to the other (a–b). Indeed, many young animals act out in play the competitions that will later be in earnest. Two young cockerels, for example, stalking and attacking each other, then turning aside to peck at the ground in an apparently irrelevant manner, frequently show the behaviour suggested by the region of Figure 20 around the origin of the pleat. To account for this behaviour, some existing theories borrow the concept of 'displacement' behaviour from Freud's human psychology; the catastrophe model presented here does not require that essentially *ad hoc* assumption, because the control factors interact in a way that makes the behaviour in part of the surface subject to random fluctuations in the level of motivation, such as distraction by a promising-looking pebble.

AGGRESSION AND MOTIVATION

Zeeman's model of aggressive behaviour in the dog is one of the most widely known and heavily criticized applications of catastrophe theory. Zeeman began with Konrad Lorenz's observation that a dog's facial expression reflects its anger or fear. Anger, Lorenz said, could be correlated with the extent to which the dog's teeth were bared, and fear with the extent to which its ears were flattened against its skull (presumably, as Darwin believed, to make the ears less vulnerable to an enemy's teeth). Although the two drives or motivations are conflicting, if they are present at high levels they do not cancel each other out. A dog experiencing neither anger nor fear is likely to behave in some 'neutral' way, but a dog experiencing high levels of both drives is likely either to attack or to flee.

The model Zeeman proposed is intended to reflect this bimodality of behaviour and two other observations: first, that it is very difficult to predict which mode of behaviour will appear when both influences increase at once (as many of us know from experience), and second, that the transitions between aggressive and submissive behaviour are discontinuous. When a dog is losing a fight with another, it does not fight less fiercely, then become neutral, then show increasing submission. Instead, it breaks off the fight suddenly and either flees or rolls on to its back (a submissive display that acts to dampen its opponent's aggression).

Figure 21 shows Zeeman's model, with anger and fear as control factors and the most likely level of aggression as the

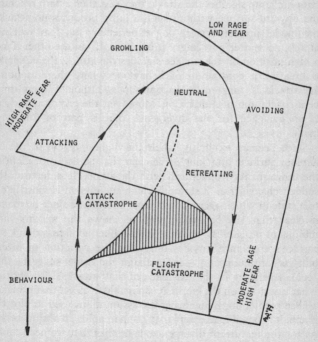

Figure 21. Aggressive behaviour in a dog (NOTE: *The control axes are oriented as in Figure 9.*)

resultant behaviour. Anger alone causes the dog to attack, and fear alone causes it to submit or flee. Neutral behaviour is most likely at low levels of both, but very unlikely when both are strong. The path between aggressive and submissive states can be either smooth or discontinuous, and the path from low levels to high levels of both drives passes near the singularity, so that a slight preponderance of one drive or the other can make a large difference in the most likely behavioural outcome.

Sussmann and Zahler contend that Zeeman's model is based on an ambiguous definition of 'attack'. If the upward catastrophic jump signifies the attack, they say, then a path reaching the upward surface without such a jump produces no attack: the model predicts that a dog experiencing no fear will not attack, no matter how angry it becomes. If, on the other hand, a high position on the surface signifies the attack, then Zeeman is creating a continuum of behaviour where there is none; an attack is *inherently* discontinuous, although there may be a continuum of motivation. Moreover, the critics add, there is no evidence for the divergence that is part of Zeeman's model.

Since Zeeman explicitly labels the highest region of the behaviour surface 'attacking', it is clear that he does not identify the upward catastrophe itself with the attack. It is, instead, the sudden change from a state in which submission is most likely to a state in which aggression is most likely. And since he makes it clear that the behaviour surface shows the maxima and minima of *probability* (what he calls a 'likelihood function'), the surface's continuity implies only that the variation in probability is continuous, not – as Sussmann and Zahler suggest – that Zeeman believes in the existence of 'half-attacks'. By ignoring what is written on the graph, it is possible to assert that it makes a false prediction; by ignoring the assumptions used to plot the graph, it is possible to assert that Zeeman holds an absurd belief. As for the feature of divergence, it implies that we need much more knowledge of a dog to predict its most likely behaviour when both anger and fear are increasing than we do to predict its most likely behaviour when only one factor is increasing. Anyone who is more cautious in approaching a stray dog than a familiar one has probably had the 'experimental evidence' for divergence which Zeeman's critics believe is lacking.

GROUP FORMATION

We often distinguish between 'social' and 'solitary' animal species as if these two modes of behaviour were genetically fixed, that is, as if all species must conform to either one or the other type. Throughout recorded history, however, at least one animal – the locust – has demonstrated periodically that transitions from one mode to the other are possible. For many years on end, locusts are unobtrusive creatures. Then, for reasons we have only recently begun to understand, they gather in swarms that devastate grasslands and crops in a geographic belt extending from central Africa to India. Because locust 'plagues' have such an economic impact, any theory that helps us understand their formation is of special interest. Catastrophe theory can do so, and it can indicate qualitatively how swarming might be prevented.

The behavioural characteristics of the locusts are reflected in their names: the *solitaria* type; the *gregaria* type, which swarms; and the *transiens* type, which exhibits intermediate behaviour. All three types are genetically identical, but the expression of the genes (what biologists call the 'phenotype' as opposed to the 'genotype') differs. *Solitaria* locusts appear to prefer bare ground for egg-laying, and have little impact on even sparse vegetation. The swarming *gregaria* locusts, on the other hand, require thick vegetation to support themselves.

The transformation from *solitaria* to *gregaria* appears to depend in part on the locusts' population density. When the concentration of locusts in a given area increases, the females lay heavier eggs. Locusts hatched from these eggs are more active and form social groupings, thus increasing the population density even further. Another factor in the transformation is the airborne concentration of a pheromone – an organic chemical that acts as a chemical messenger outside the body, rather than inside the body as is the case with hormones. The locust pheromone is produced chiefly by immature locusts. It stimulates maturation and may also preserve the integrity of swarms (once they are formed) by acting as an attractor.

Figure 22 illustrates how population density and pheromone

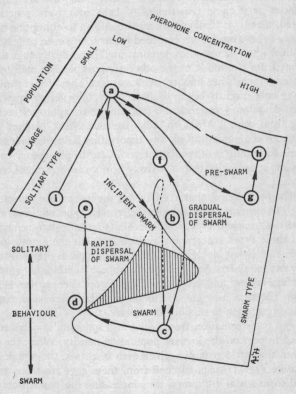

Figure 22. The swarming behaviour of locusts

concentration may interact to control locust behaviour. Initially (*a*), both control factors are at low levels. As the population increases through successive breeding cycles, the pheromone concentration also increases. At a critical level of both (*b*), the *transiens* locusts change swiftly to *gregaria* and a swarm erupts (*b–c*).

However, there are several external influences which oppose the stability of a swarm – fortunately for farmers. If a swarm does not migrate fast enough, it can deplete the available food supply. Also, actively swarming locusts tend to spend less time in breeding, so the proportion of immature locusts in the swarm decreases. This lowers the pheromone concentration, and the swarm loses its integrity. It can break up rapidly (*c–d–e*) or, if starvation intervenes to cut the population, in a gradual disintegration (*c–f–a*).

Catastrophe theory's qualitative analysis suggests several ways in which locust swarming might be controlled. For example, if the pheromone could be synthesized and sprayed into the air around an incipient swarm, it would maintain the insects in an artificial *transiens* form (*a–g*). The pheromone would be effective in much smaller quantities than insecticide, and it would gain time for changing weather (or, if necessary, insecticide) to reduce the locust population (*g–h–a*).

Alternately, an artificial *lowering* of the pheromone concentration would also prevent swarming. This lowering could be achieved by finding a substance that neutralizes the pheromone in the air, or a substance that blocks its metabolic effects on the *transiens* locusts. Another method would be to breed and release the albino mutant locust, which does not produce increased pheromone when crowded. Either of these methods would allow population increase, but prevent the transformation to *gregaria* (*a–i*).

Catastrophe theory, of course, had nothing to do with the discovery of the locust pheromone. Undoubtedly experiments in locust control via pheromones would have been conceived without it. But without the catastrophe theory model to clarify the interaction of pheromone concentration and population density, the results of such experiments would be harder to interpret. For example, a researcher observing that a very small change in pheromone concentration had a large divergent effect on behaviour might conclude that there was a threshold level of pheromone effect. In the perspective provided by this model, however, it is clear that the pheromone's effects vary continuously with its concentration. It is the *combination* of pheromone level and population density that produces the apparent threshold.

Catastrophe theory neither contradicts nor replaces the approaches to animal behaviour described at the beginning of this chapter. Rather, it is compatible with all of them, and provides a new way to organize their assumptions and observations. The models proposed in this chapter use only the properties of the cusp catastrophe surface – bimodality, catastrophe, hysteresis and divergence – and they reflect only the behaviour of a single species. To show how different species interact behaviourally, or how these interactions change with time, more complex models based on higher-dimensional catastrophe surfaces, such as the butterfly, are needed. And beyond the elementary catastrophes, there is Thom's vision of a dynamic, global model for all living systems. In that model, the behaviour of each species would be a small part of the unfolding of a singularity of unimaginably high dimension. Two species that are dependent on each other (bees and snapdragons, for example) would be topologically 'closer' to each other than two species which have many genes in common. Perhaps Thom was moved to this thought by feelings similar to Darwin's when, at the end of *The Origin of Species*, he passed from the linear imagery of the branching evolutionary 'tree' to a more complex vision:

It is interesting to contemplate a tangled bank, clothed with many plants of many kinds, with birds singing on the bushes, with various insects flitting about, and with worms crawling through the damp earth, and to reflect that these elaborately constructed forms, so different from each other, and dependent on each other in so complex a manner, have all been produced by laws acting around us ... whilst this planet has gone cycling on according to the fixed law of gravity ... endless forms most beautiful and most wonderful have been, and are being evolved.

7

APPLICATIONS IN SOCIOLOGY
AND ECONOMICS

> 'This [authority] function u must have at least one
> maximum, and the individual at this point is the
> chief, since he takes orders from no one' – RENÉ
> THOM

THE social or 'inexact' sciences have an uneasy relationship
with mathematics. To some extent, they seek a Newtonian
goal of quantification and prediction. Yet the human and environmental variables they must deal with are so many and
varied, the possibility of meaningful experiment so limited, and
the data (both current and historical) so questionable, that the
greatest achievements of sociology and economics so far are
chiefly descriptive rather than analytic. In addition, any theory
in social science faces a special problem : the widespread fear
that if knowledge is power, then knowledge in the social
sciences could reinforce the power of those who may already
have too much. There is a justifiable concern about the increased threat to privacy and freedom posed by government
as it gathers data to feed the statistical and econometric models,
and makes decisions based on computer extrapolations of current trends.

The most widely used mathematical tools in the social sciences
are statistical, and the prevalence of statistical methods has
given rise to theories so abstract and so hugely complicated
that they seem a discipline in themselves, divorced from the
world outside learned journals. Statistical theories usually
assume that the behaviour of large numbers of people is a
smooth, average 'summing-up' behaviour over a long period of
time. It is difficult for them to take into account the sudden,
critical points of important qualitative change. The statistical

approach leads to models that emphasize the quantitative conditions needed for equilibrium – a balance of wages and prices, say, or of imports and exports. These models are ill suited to describe qualitative change and social discontinuity, and it is here that catastrophe theory may be especially helpful.

Two of the models suggested in this chapter are purely descriptive, intended to illustrate what is already known about social psychology and the workings of social status; two more, in economics, indicate features that might be looked for in present and future situations, perhaps with an eye to policy-making and prediction.

SOCIAL PSYCHOLOGY: CROWDS AND ARMIES

The unpredictable behaviour of large groups at critical moments of stress has done much to shape history. The revolutionary crowd that stormed the Bastille in 1789 was coherent and purposeful at such a critical moment. The prized Imperial Guard of Napoleon, as it broke under fire at Waterloo in 1815, lost its vital cohesion under supreme stress. In 1895 Gustave Le Bon theorized in *La Psychologie des foules* (*The Psychology of Crowds*) that what happened at moments like these was a manifestation of a 'crowd mind', of the 'immense unconscious working' of 'mysterious forces'.

Social psychology has not advanced much since Le Bon's time in understanding this kind of behaviour. A model based on the cusp catastrophe surface may help us to organize and understand more readily what we know. Figure 23 shows the model, with the orderliness of group action as the behaviour axis. The control factors selected here are cohesiveness, the tendency for individuals to identify with their group and its aims, and danger (actually the perceived level of danger, since a rumour can be at least as effective as a solid fact in determining group behaviour).

Path *a–b* on the surface accords with the observation by many commanders and military historians that a well-trained army, in which the soldiers are conditioned to regard the in-

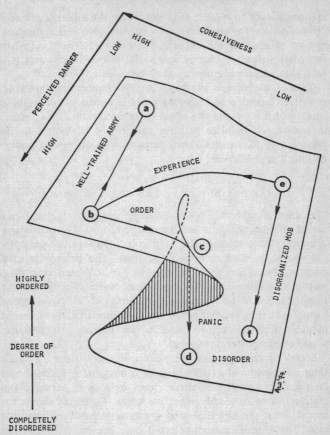

Figure 23. *Social order v. disorder in times of danger*

tegrity of their unit – from a squad to a division – as of para-
mount importance, become even more closely co-ordinated as
danger increases. If its cohesiveness decreases, for example, if
the soldiers see comrades fleeing the field, the breakdown of
the army may be sudden (*b–c–d*).

A mob, unlike an army, becomes less orderly with increasing danger (e–f). But if its sense of cohesiveness increases steadily while the danger rises, the ground can 'turn the corner' on to the upper surface (e–b). This path corresponds to the development of the Chinese Communist forces on the 'Long March' of 1935, which turned an initially ill-organized group into the core of the force which would eventually overwhelm the Nationalists in 1949.

This model makes it easy to see why rumours can be so demoralizing. In addition to heightening the perceived danger, they lower cohesiveness by hinting that the 'official version' of events is false – in other words, that the leaders don't even trust their followers with the truth. The model also suggests that even when news from official sources is believed, it can lead to panic and a sudden loss of order. The 1938 broadcast of *The War of the Worlds*, for example, first heightened the perceived danger with a vivid description of the non-existent Martians' landing and attack, then reduced cohesiveness by breaking up into confused 'reports' of the army's retreat, the roads choked with refugees, and so on. Those who heard the broadcast felt all at once that there was nothing between them and the alien invaders, and panic ensued.

The only 'prediction' of this model is something that every successful leader knows: that when the level of danger is about to increase, even a small increase in cohesiveness can make a great deal of difference. Shakespeare showed this unforgettably in Prince Hal's incognito visit to his soldiers' fireside the night before battle. Although they do not recognize him and much of the scene is played for comedy, the audience is left with an impression of closeness between Henry V and his motley soldiers, and so is instinctively prepared for the coming victory.

STATUS AND MARRIAGE

Another purely descriptive model is that shown in Figure 24, relating social status to its two most important determinants (at least in many Western societies): wealth, and the less

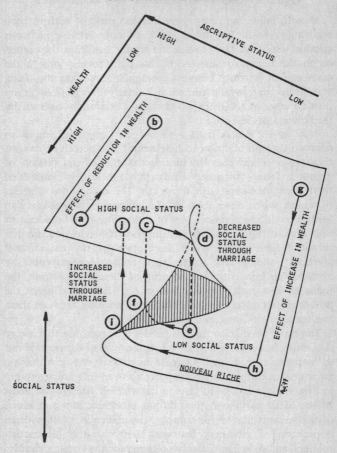

Figure 24. Smooth and sudden changes in social status

measurable attribute sociologists call 'ascriptive status', which is obtained through family background, occupation, philanthropy or 'breeding'.

What do the various positions and paths on this model surface correspond to? Those fortunate enough to enjoy both wealth and high ascriptive status, of course, have high social status (a): a seat on the stock exchange and a seat at the opera are the best credentials in this region. Even high ascriptive status alone can still ensure a reasonably high social status (b) – that of a poor but respectable college professor, for example! The path from a to b is a gradual decline, that which leads to the faded gentility of old families in 'reduced circumstances'.

In male-dominated societies the social standing of the wife generally follows that of her husband, so that the effects of marriage on status can be represented as jumps on the model surface. A low-status woman who 'makes a good match' – Cinderella, for example – makes the upward transition e–f–c, while the reverse transition c–d–e (marrying 'beneath oneself') poses a threat to the status of the woman involved, but not to that of Prince Charming. The model also incorporates the seemingly paradoxical decline in status of the *nouveaux riches* (g–h), whose increasing wealth alone makes them more vulgar (in the social establishment's eyes) than they were before. Because of divergence, the best course for one who is beginning to succeed financially is to concentrate on good manners, good diction, good works and the other aspects of ascriptive status. These are the signs his 'betters' will scrutinize before allowing him on to the upper sheet of the surface. The daughters of the *nouveaux riches* can easily attain high status via marriage, leaving their kin behind (h–i–j).

This simple model reflects the changes of individuals' social status, but it might be adapted to follow the shifting relationships of whole social classes in history. The emergent layer of intermediate behaviour in the butterfly catastrophe, for example, could correspond to the development of the bourgeois 'middle class' that plays such a large part in sociologically oriented views of history. If the appropriate control factors in such a model were found, we might be able to understand why the rise to social dominance of a middle class is sometimes gradual and sometimes marked by revolution and civil war.

ECONOMICS: COMPETITION AND PRICES

One striking weakness of existing economic theory is that it gives a clear picture of the behaviour of prices in either a competitive or a monopolistic situation, but is much less clear about what happens as one kind of situation changes towards the other. A description based on catastrophe theory may provide a model for these transitions where none existed before. It should stimulate economists to seek new data to verify the model's qualitative prediction : that under certain circumstances, two distinct price ranges for a given commodity should exist as the number of producers of that commodity changes.

First, a few basic definitions. The number of producers in an industry may be determined by its economies of scale. If production is cheaper in large quantities, the industry tends to be dominated by one company (a monopoly) or by a few (an oligopoly). The supply of electricity to most American cities, for instance, is provided by monopolistic utility companies, while the major automobile manufacturers form an oligopoly. In each case the circumstances of production promote a 'natural' monopoly or oligopoly.

In an oligopoly, the production level of one firm has a great impact on the prices, sales and profits of the other firms. This impact can be reduced by the adoption of one of three strategies. First, the oligopoly can form a cartel in which the production and prices of each member are determined by negotiation. In effect, the oligopoly then behaves as if it were a monopoly. Second, one or more firms can initiate a round of predatory price-cutting to force weaker members out of the market. Third, in some circumstances firms can merge, reducing the level of competition to the benefit of the remaining members.

Even when these strategies are employed, though, manufacturers are generally unable to raise prices indefinitely. As prices go up, customers increasingly look to alternate sources of supply or alternate ways of fulfilling their requirements, or simply stop buying altogether. The extent of this behaviour is reflected by an economic quantity known as elasticity of demand. Low elasticity of demand is characteristic of the commodities generally considered to be necessities : even if the price increases sharply, there is only a small decline in demand. High elasticity of demand is associated with goods and services whose purchase can be delayed or omitted altogether : a small increase in price causes a relatively large decrease in demand. Customers will still buy some of the commodity even at a high price, but they will buy much more at a low price. In this case, producers tend to be more competitive in price.

So both the number of producers and the elasticity of demand can affect prices, and a complete model of their influence should show both gradual and sudden changes in price. There are mathematical models for both kinds of change, but there is no single model incorporating both. Figure 25 is a first step towards remedying that situation. Under conditions of low elasticity of demand and high competition, the price of a commodity is relatively low (a). If the number of producers decreases while elasticity of demand stays low, the price increases slightly but the market remains basically competitive (a–b). A further reduction in the level of competition, though, may bring about the formation of an oligopoly and a dramatic increase in price (b–c–d). Mergers or cartelization can create an actual or de facto monopoly (d–e) that charges even higher prices.

An example in which the formation of an oligopoly had a dramatic impact on price was the establishment of OPEC, the association of oil producing and exporting countries. These nations had previously sold their oil competitively, but in 1973 they began to set prices in concert as an oligopoly. The price of crude oil went from $2·12 per barrel in January 1973 to $7·61 per barrel a year later, and to $10·50 per barrel by January 1975.

This price increase did have some effect on consumption, which fell by 14 per cent in Belgium and the Netherlands, 10 per cent in West Germany and 3·5 per cent in the United States.

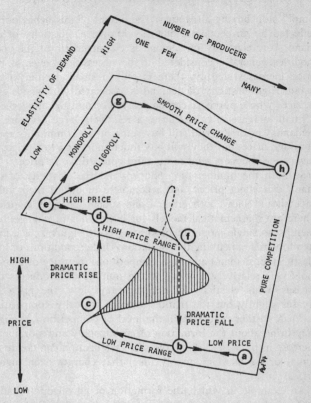

Figure 25. Effects of competition and elasticity of demand on prices

But the elasticity of demand for oil, in the short run at least, was low, and the members of OPEC had little difficuly in selling all they chose to produce.

The level of competition could increase in any of several ways. If one or more members of OPEC should break with it and produce much more oil at a low price, they would be able to increase their relative share of the world market. Or the high price could encourage the location and development of new sources of oil, for example, the North Sea fields or those in Alaska. In either case, one would expect that a sufficient increase in the number of producers would bring about a dramatic price reduction (d–f–b). Naturally, potential new producers realize that this reduction is likely to be delayed well beyond the level of competition at which the oligopoly first emerged. This is an obvious feature of the model.

What if the elasticity of demand were to change – if, for instance, an alternate source of energy were to become readily available? A total monopoly can withstand such an increase in the elasticity of demand, though at least some price decline is inevitable (e–g). An oligopoly experiences greater pressure as its members begin to disagree over the price cuts needed to remain competitive in the new situation, and a gradual disintegration may take place (e–h) with a large but relatively smooth price decline.

In cases of very high elasticity of demand, the catastrophe model is compatible with traditional analysis in predicting a smooth change in prices as a market shifts between conditions of low and high competition (g–h). In this region of the surface, there is little incentive to form a cartel, because the higher prices it permits would be met by a decline in demand. In addition, the model indicates that the actual difference in price between competitive and monopolistic situations should be less than it would be if the elasticity of demand were low. (It is assumed in economic theory that if demand were perfectly elastic – that is, if any increase in price were to eliminate demand entirely – there could be no change in price at all. The catastrophe surface does not have a completely flat region, but, of course, perfectly elastic demand never really occurs, either.)

The most interesting feature of this model is its prediction of two price ranges, one high and one low, in conditions of low to moderate elasticity of demand and moderate competition. As we have seen, the transitions between these ranges would be dramatic in cases of very low elasticity of demand. They would be expected to become less pronounced as elasticity of demand increased: a 'boom or bust' cycle would grow smaller, because the incentive to take advantage of high prices – or leave the market because of low prices – would grow less.

A concomitant feature is that the price range in effect at a given time depends on whether the market's level of competition has been increasing or decreasing. Until the transition points are reached, high (but gradually declining) prices will be associated with increasing competition, and low (but gradually increasing) prices with declining competition. This qualitative prediction does not arise in existing economic theories. It should be possible by searching existing records or by gathering new data to determine whether it holds true in practice, and perhaps to find the quantitative dimensions of this 'bimodal pricing' in specific cases.

A more complex model, one based on the butterfly catastrophe surface, would be needed in order to describe the interaction of monopolies or oligopolies. If the producers involved in an oil cartel, for example, found that the producers of grain, or chromium, or another vital commodity had formed their own cartel, the cusp catastrophe model would not be of adequate dimension – but a butterfly model's intermediate region could depict the compromise that the cartels might attain. This kind of qualitative refinement of the model will certainly be necessary to build any theory for the many-tiered, interlocking price agreements that dominate much of international trade today.

INFLATION AND EXPECTATION

Economics as a predictive science is inevitably bound up with politics, especially where levels of inflation and unemployment are involved. Empirical studies have suggested that there is a trade-off between these levels, and political parties accuse each other of sacrificing those out of work to the goal of lowered inflation, or conversely of sacrificing economic stability to the goal of full employment.

More recently, both economists and politicians have recognized that the *expectation* of future inflation is an important factor. If a high level of inflation is the norm, then workers begin to demand higher wages to offset the increased cost of living they believe will come during the period of a contract. These higher wages themselves have an inflationary impact, especially when they are attained by large numbers of workers at once through the negotiations of major unions.

This phenomenon suggests that a qualitative model for inflation should include the expected inflation rate as one control factor and unemployment as the other. Figure 26 shows such a model. The worst case (in terms of its effects on inflation) is low unemployment and a high expected inflation rate (*a*). It can be improved somewhat by lowering the expected rate (*a–b*); this may be achieved by a government's adoption of an aggressive policy to discourage price increases. To achieve a greater decline in inflation, it may be necessary to permit a politically unpopular increase in unemployment at the same time (*a–c–d*). Increasing unemployment alone, with no decrease in the expectation of future inflation, will produce only a slight decline in inflation (*a–e*). Conditions in the region around *e* correspond to the 'stagflation' (stagnant inflation) that has plagued Great Britain and Italy from the early 1970s until the present. To move from this region to one of lower inflation would require a drastic increase in unemployment (*e–f–g*) or, preferably, a slight increase in unemployment coupled with credible steps to reduce future inflation (*e–c*). A knowledge of the precise state of the economy around *c* is obviously crucial, since an increase in expected inflation from that point can have widely divergent effects on the actual inflation rate.

Many of the debates over economic policy in recent years have concerned what is described as the 'fine tuning' of quantities such as the money supply (which affects inflation via loan rates) and the government's expenditure for goods and services (which increases employment by raising total demand). This

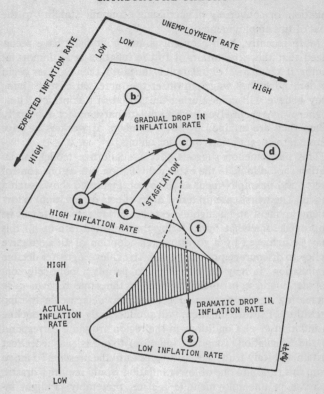

Figure 26. Gradual and dramatic changes in inflation

model's value lies in its indication that the *sequence* in which the control factors are altered can, at any given moment, be at least as important as their quantitative levels.

Some of the most vehement criticism of applied catastrophe theory has been directed at models in economics and sociology. A model developed by Zeeman for stock-market crashes, for example, has been interpreted by Sussmann as implying that a purely speculative stock market could never experience a crash, which in Sussmann's view contradicts both reason and experience. The assumptions behind Sussmann's charge appear to be the same as those prompting his criticism of the model for aggression in dogs : that a discontinuous change in the level of some type of behaviour is indistinguishable from a rapid but continuous change.

Another kind of criticism, inspired by the fear of Big Brother, was directed at the model for a prison riot developed by Zeeman and several prison psychologists. That model was retrospective rather than predictive, since it was worked out in 1975 on the basis of data gathered in 1972. The data were entirely 'public' – that is, they concerned the numbers of prisoners petitioning to see the prison governor (warden), the numbers asking for transfer out of a given cell block, etc. The paper presenting the model made only one 'prediction', namely, that prison authorities would do better to deal with a confrontation by negotiation than by the use of force. Yet Jonathan Rosenhead's criticism overlooked all of these points, claiming that catastrophe theory was being applied 'not for liberation but for social control. Another management tool is added to the existing battery of science-based techniques for keeping prison populations subjugated – behaviour modification, aversion therapy ... psychosurgery, riot gases and impact weapons.' There is nothing in Zeeman's model itself, or in its very tentative presentation, to justify such a diatribe. Even an accidental misunderstanding sparked a similar response : a magazine atricle mistakenly reporting that cameras in prisoners' cells were part of Zeeman's data-gathering process led to an inquiry in Parliament before the mistake was corrected.

In fact, catastrophe theory has none of the ideological leanings that some of its opponents see in it. Its descriptions of continuous and discontinuous change are not weighted in favour of 'social control' or against 'liberation'. Instead, they offer the same qualitative insights to anyone (of whatever political or social persuasion) who chooses to take advantage of them. The theory of evolution was used to justify ideas of ruthless social and economic competition in the late nineteenth century, but that misuse did not invalidate the theory. Catastrophe theory has scarcely been used at all, and it seems much too early to claim that it supports any one particular view of the social world.

8

APPLICATIONS IN POLITICS
AND PUBLIC OPINION

'It seems that the social mind has a fragmentary
character very similar to that of the animal mind:
society finds its identity only in the face of an
urgent threat, like war, where its existence and
stability are threatened ...' – RENÉ THOM

HOW we view political developments is very much a matter
of time. The historian has leisure in which to analyse events of
the more or less distant past, although the luxury of historical
hindsight is balanced by the difficulties of assessing what people
many years before actually did, thought and intended. The
political journalist, dealing with current events, has far less
leisure and far less perspective in time, but can observe day-by-
day developments in all their detail and talk to those involved.
And the political scientist, drawing on material from history and
current events alike, must attempt to deduce general principles
applicable to both past and present.

If political science were able to make detailed predictions of
events, then there would be few political scientists in our universities : they could be far more influential (not to say prosperous!) working for political leaders or organizations. In fact,
we speak of the 'art of politics' more readily than of political
science. Most practical politicians are, if not artists, engineers.
They have their own rules of thumb, working experience in the
field, and pragmatic tests (such as public opinion polls and
elections) to tell them what works and what doesn't. And for the
most part, they have little use for abstract political theories.

Catastrophe theory may be useful in political science in three
ways. First, because of its independence of time scale, it can
describe processes that occurred over many centuries as readily

as those detailed in this week's newspapers. Second, its organization of control factors and behaviour provides a link between
quantitative data (the cost of a war, say, or the results of a
poll) and qualitative changes in behaviour – those which are
most likely to perplex the historian and catch the politician off-
guard. Third, because of its distinction between continuous and
discontinuous change, it offers a picture that combines political
evolution, the more or less continuous processes we call 'trends',
with political *revolution* (literal or figurative), the more or less
discontinuous events that separate one period of political development from the next. The applications of catastrophe theory
presented in this chapter deal with the history of Rome, with
the quantitative causes and qualitative effects currently involved
in public opinion on nuclear-power policy, and with the evolu-
tions and revolutions that can lead from democracy to dictatorship and vice versa.

THE ROMAN EMPIRE: CHALLENGE AND RESPONSE

One of the most widely influential historical ideas is expressed
in Toynbee's *A Study of History*, where the course of great
civilizations is charted in terms of challenge and response. An
external threat, Toynbee concludes, can either enhance or
weaken a civilization's integrity. As a result, that civilization
may grow to dominion over surrounding cultures, or be subjugated by them.

The early growth of Roman power, for example, was accel-
erated by Rome's conflicts with a series of opponents: first
with the Etruscans and with leagues of other Italian city-states,
then with the declining Greek civilization and with Carthage.
In Figure 27 these contests can be envisaged as swerves in the
line from *a* to *b*, none of them reversing the overall trend of
Roman dominance as one opponent after another gave way to
increasingly centralized Roman rule.

The Punic Wars, which ended with Carthage's destruction,
brought Rome to political supremacy in the Mediterranean in
the late second century B.C. But those wars also helped bring

about the changes that would eventually destroy the Roman republic. In the words of historian W. H. McNeill, 'Years of interminable campaigning pried many peasant soldiers loose from their ancestral farms; and an idle urban proletariat, which subsequently played an important political role, began to drift into Rome. Simultaneously, senators and tax farmers who collected provincial revenues became rich beyond all Roman precedent.'

Slave labourers became increasingly common as large plantations replaced the old farms and agricultural productivity declined. Competition from new industries, principally in Gaul, became acute after 50 B.C. The cost of imports to Italy had to be met with precious metals, mostly won as plunder from Roman conquests. When that was not enough, the large landowners were taxed to pay for the armies, and their economic and political power was transferred to the emperors. Augustus effectively ended the Roman republic in the last generation B.C. Though the frontiers of Roman power continued to expand, the integrity of Roman civilization was waning. Eastern religions such as Christianity grew in popularity, while the armies lost their allegiance to Rome as a political ideal and became instruments for ambitious generals who sought the imperial title.

Rome was still able to meet external challenges in the short run. It held off the Sassanid rulers of Persia, for example, in the third and fourth centuries A.D. But to do so it compromised the empire's integrity by granting its eastern client states more independence and freedom of action, in effect creating a 'buffer zone' between Roman and Persian power. This process is illustrated in Figure 27 as path *b–c–d*. Over the long run, though, the decline made catastrophe inevitable. The fall of Rome came when peoples who had been Roman subjects turned inwards against their former rulers. It was precipitated by a change in military technique, the development of armoured cavalry which overwhelmed the Roman infantry legions and their light cavalry. The new conquerors were products of the heavy, well-watered soils beyond the Mediterranean fringe, where abundant fodder could be had. Roman agriculture, long neglected, could not support both the swollen urban population and the necessary

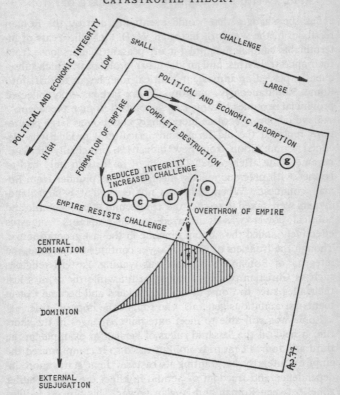

Figure 27. The rise and fall of the Roman Empire

cavalry. The Huns shook the empire, and the Visigoths and Germanic invaders shattered it (*d–e–f*).

Centralized power survived in the eastern Mediterranean, where Constantinople (Byzantium) resisted the invaders from behind its walls and its fleets for a millennium. But the western portion of the empire was completely destroyed (*f–a*), to be replaced by transient, petty states under Visigothic or Frankish chieftains. The Iberian peninsula, for example, was an easy conquest for the expanding Moslem power in the early eighth century (a–g).

POLITICAL INVOLVEMENT AND CONTROL

In modern times, we have experienced revolutions and civil wars as destructive and sudden as any conflict between rival empires. The civil war that followed the Russian Revolution, for example, was more costly in human life than Russia's campaigns during the First World War; and Hitler's rise to power brought swifter and more radical change to Germany than had the end of the Second Reich in 1918. Catastrophe theory can help us see how political factors interact to produce such events.

To model the changes in political activity, we can use as control factors the degree of popular involvement and the degree of central political control (Figure 28). Where central control is not overwhelming, a change in the level of popular involvement does not bring about political upheaval (a–c): indeed, many political parties may coexist, sharing power in coalitions that appeal to a broad public.

Gradual political evolution, with increasing central control, can lead either from an oligarchy to a personal dictatorship (c–b) or from a democracy to a 'dictatorship of the proletariat' (a–e). But if central control is already high, a pronounced increase or decrease in popular involvement is likely to produce a revolution (b–d–e or e–f–b). If popular involvement increases and then decreases, a democratic trend may be followed by counterrevolution. This was the case, for example, in Czechoslovakia in 1968, when the 'Prague Spring' of the Dubček government was ended by Russian intervention which re-established strong control and suppressed popular involvement (b–a–e–f). In India the brief near-dictatorship of Indira Gandhi reflected a popular disillusion with democracy, allowing an elected ruler to assume concentrated power (a–b), followed by a successful opposition movement and a surprising election in 1977 (b–a). This non-violent transition can occur in conditions of moderate control, where the catastrophe surface is folded over itself but

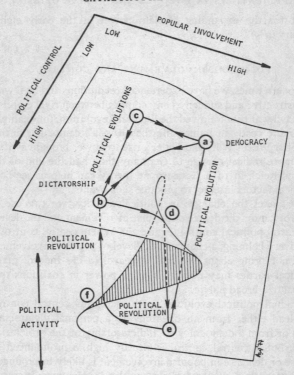

Figure 28. *Political evolution and revolution*

the transitions are not great enough to be revolutionary. It is here that we would expect to find the marked but peaceful shifts that can occur in two-party systems. Traditionally, one party identifies itself with the 'establishment's' power structure, while the other is the 'people's' advocate. Prosperous times, when the political pie is big enough to satisfy everyone, favour the establishment because popular involvement declines into sentiments such as 'you never had it so good' and 'don't rock the boat'. Hard times, on the other hand, lead people to adopt more militant political stances, and the dominant slogan is 'it's time for a change'.

Interestingly, the overlap in this region shows up at times when the ostensibly conservative party argues against 'big government', or the ostensibly liberal one advocates broad regulatory power for centralized agencies. An example of this seeming paradox occurred not long ago in Great Britain, where the Conservatives under Prime Minister Heath adopted increasingly socialist measures, while Harold Wilson's Labour Party opposition, socialist in theory, became more conservative. For a time it was commonplace to remark that the Conservatives were more socialist than the socialists!

The qualitative features of this model suggest intriguing speculations. A democratic government, for example, could 'keep track' of its relative position on the catastrophe surface through polls and other measuring techniques, maintaining power by manipulating events to avoid the fold line of the surface. A revolutionary organization, on the other hand, attempts to move society towards the discontinuous transition it advocates. Thus today's 'urban guerrillas' and terrorists can be seen as provoking a higher level of political control by their bombings and hijackings, moving the point representing society to the region of potential instability. In the events leading up to the Russian Revolution, Lenin is said to have been very concerned about the precise timing of his Bolsheviks' moves towards political power. In the right circumstances, he argued, only a small 'push' would be needed to bring about a transition that at other times would need a much greater effort. The relative ease with which the Bolsheviks displaced the supposedly democratic Kerensky, seven months after the initial phase of the revolution had overthrown the Tsar, indicates that Lenin was very successful at judging when to apply that push.

ISSUES AND INTERESTS: CONFLICTING LOBBIES

The conflict between two organized pressure groups with opposed aims can be modelled as shown in Figure 29. From the

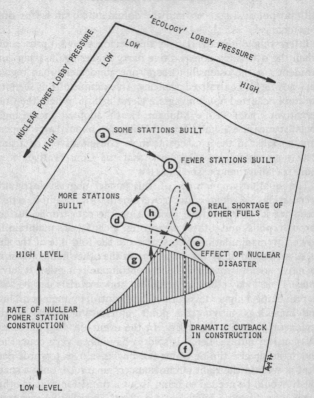

Figure 29. Conflicting lobbies: ecology v. nuclear power

mid-1950s until the beginning of the 1970s, a combination of
private initiative and encouragement by the Atomic Energy
Commission (which acted in effect as an 'internal lobby' for
nuclear power) led to the construction of a number of nuclear
power stations (*a*). The rise of the environmental movement,
and the formation of an 'ecology' lobby, has led to the current
slowdown (*a–b*). As today's arguments suggest, the US is at a

critical turning point. Increased oil prices and the desire for ever-increasing amounts of energy are pressures favouring the construction of many additional nuclear power stations in the coming generation (b–d), while fears of irreparable harm to the environment work in the opposite direction. Proponents of each side believe that the other side will eventually see the light. Nuclear-power advocates claim that a real shortage of other fuels must eventually weigh more heavily than ecological arguments, and a crash programme of construction will be needed (c–g–h). Spokesmen for the ecology lobby foresee that more nuclear-power stations will inevitably hasten the occurrence of a nuclear disaster and a consequent reaction (d–e–f).

Figure 30, which changes the model from a cusp to a butterfly, gives an alternate picture that shows how a compromise might emerge. One change favouring such a compromise is the Atomic Energy Commission's absorption by the recently established Energy Research and Development Agency and the Department of Energy. The stated objective of this change was to establish comprehensive national assessment of energy needs, and to integrate government plans to meet those needs. At almost the same time, the creation of the Environmental Protection Administration has produced an 'internal lobby' for ecologically sound policies. While perfect harmony among government agencies is no doubt impossible, they must now deal with each other on a daily basis, which seems to offer a better chance of compromise than the periodic confrontations of nuclear-power and ecology lobbyists outside the government.

Another changing factor that may work to encourage compromise is the increasing emphasis on energy sources other than nuclear power or imported petroleum. If the United States' extensive coal reserves can be exploited in an ecologically acceptable way, or if solar power generation should become more economically attractive through improved technology, the urgency of nuclear-power station construction would be lessened, and its advocates could concentrate on those situations where it is clearly superior to the alternatives.

Neither of these important changes was visible at the begin-

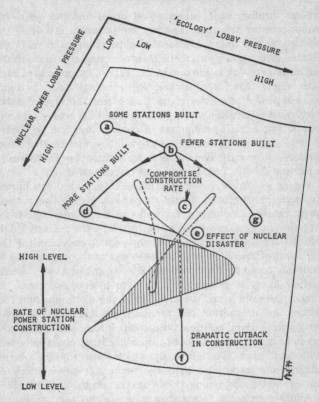

Figure 30. The emergence of compromise

ning of this decade, but together they may spare us the hard choice of many nuclear-power stations or none at all. Their influence opens up the compromise path *b–c* in Figure 30. We often say that an unexpected development 'adds a new dimension' to a political situation; in this model based on catastrophe theory, that figure of speech becomes literally true. In changing from the cusp to the butterfly, we take into account additional factors which create the possibility of compromise : the intermediate sheet of the butterfly behaviour surface. This might be what mediators call 'common ground', a region where groups with polarized, conflicting interests can (however grudgingly) arrive at a consensus.

The application of catastrophe theory in political science has been attacked by Sussmann and Zahler on grounds similar to those on which they criticize applications in the other social sciences. For example, they charge that just as Zeeman falsely postulates a continuous range of aggressive behaviour in his model for the dog, so he and Carlos Isnard falsely postulate a continuous range of military activity in their model for public opinion (which relates the perceived cost and threat of war as control factors). The Zeeman–Isnard model suggests that as the cost of the war increases, public opinion becomes increasingly split between 'hawks' and 'doves'. The critics argue that to go to war is inherently a discontinuous choice, and that there is no evidence that increasing cost acts as a divisive influence on public opinion.

A different point of view has been expressed by Robert T. Holt, a professor of political science at the University of Minnesota. 'Zeeman's papers are clues, not proofs,' he says, 'and we should use them in that spirit.' He suggests that the outbreak of the First World War, which took place after a long period of international tension, shifting alliances and threatening 'incidents' in the Balkans, might be usefully represented as a catastrophe. The war's ending, brought on by simultaneous Allied advances and political revolution within Germany, was also a discontinuous change occurring after more than four years of sustained effort by both sides had scarcely altered the lines on the Western Front.

Thom himself is both speculative and cautious on the rare occasions when he turns his theory to political affairs. 'It is tempting to see the history of nations as a sequence of catastrophies,' he wrote in *Structural Stability and Morphogenesis*; '... what better example is there of a generalized catastrophe than the disintegration of a great empire, like Alexander's! But in a subject like mankind itself, one can see only the surface of things. Heraclitus said, 'You could not discover the limits of soul, even if you travelled every road to do so; such is the depth of its form." '

9

APPLICATIONS IN
PSYCHOLOGY

*'We shall have to find the best means ... to formal-
ize the unformalizable. For this task, the human
brain, with its ancient biological heritage, its clever
approximations, its subtle aesthetic sensibility, remains and will remain irreplaceable for ages to
come'* – RENÉ THOM

THE human brain contains about ten billion cells, each with
as many as one hundred thousand connections to other cells. We
cannot possibly obtain a complete quantitative analysis of this
system, any more than we can obtain a complete quantitative
analysis of the far simpler mechanics of the solar system. In
both cases, the result is a vast array of equations that cannot be
solved in any finite time. Therefore, no useful psychological
theory can be purely quantitative.

Psychology has borrowed many ideas from physical science,
as shown by its use of words like 'stress' and 'equilibrium'. (It
would perhaps be more accurate to say that psychology has
borrowed the ideas back, since many historians of science believe that the concepts of mechanics, for instance, grew out of
our experience of muscular force and resistance.) Most psychological theories also implicitly or explicitly use the idea of
the potential by postulating (as Freud did) that our thoughts and
actions are determined by instincts seeking equilibrium in our
'maxima' of gratification and 'minima' of discomfort, or by
postulating (with Pavlov and the behaviourists) that the thresholds of response to various stimuli can be raised or lowered by
conditioning. In fact, much of psychology is the descriptive and
qualitative study of psychological equilibrium and of the stimuli
which alter it. Certainly psychological equilibrium is dynamic

rather than static, because neither instinctive pressures, nor the patterns of external stimuli, nor the electrical and chemical potentials of brain cells are at rest while we are alive. So if the descriptive and qualitative power of catastrophe theory is applied to psychology, we should think of the points on the catastrophe surface not as fixed psychological states, but as the centres of 'attractors' around which the behaviour of the brain tends to cycle. We speak of a mentally disturbed person as unbalanced, but obviously a lasting disturbance represents its own kind of balance – an equilibrium qualitatively different from that which we call normal or sane. The examples of applications in this chapter – in conditioning and behaviour, in clinical psychology and in perception – attempt to show how catastrophe theory's classification of stable states and their qualitative changes makes it a promising conceptual tool for psychology.

LEARNED HELPLESSNESS

Psychologist Martin E. Seligman has suggested that depression, the prolonged sense of 'inability to cope', is the result of a process of learning or conditioning. He calls depression a state of 'learned helplessness' which reduces our ability to respond to psychological stress. In one experiment he cites, dogs were divided into two groups: experimental and control. Dogs in the first group were subjected to electric shock while being restrained so that they could not escape. Both groups were then tested in a 'shuttle box' with two compartments, one of which was wired for shock. Ten seconds before the shock, the room lights were dimmed. With repeated trials, the dogs in the control group learned fairly quickly to move to the other compartment as soon as the lights were dimmed. The pre-treated dogs, though, typically ran about the wired compartment and then lay down and accepted the shock passively, without attempting escape. Seligman notes that they were also passive and unresponsive in other situations as well.

Learned helplessness, he asserts, has been observed in many

species, including man. Human subjects in a 1972 study by Glass and Singer were asked to perform a complex task. If a repeated loud noise interrupted their concentration, they gave up the task. But when they could control the noise by their own actions, even though it continued to interrupt them and they had to stop to turn it down each time, their skill and persistence in the task were greatly improved.

In situations like these, the avoidance response – the action taken to counteract an unpleasant stimulus – is a qualitative change in behaviour whose threshold varies with both the intensity of the stimulus and the subject's past experience. For the dogs, the response was literally a jump over the low barrier dividing the two compartments of the shuttle box; for the human beings, it was the decision to stop work and turn down the noise. Figure 31 shows this response as a discontinuous transition on the cusp catastrophe surface, with stimulus intensity as one control factor and the acquired sense of control over the stimulus as the other. The response of the 'naïve' or control group of dogs occurs along the path (a–b–c). The conditioned group, whose sense of control has been lowered, shows a reduced response (d-e-f) or, if the conditioning is sufficiently strong, an apathetic lack of response (g–h). This conditioning is reversible: when the apathetic dogs were forced to jump to the safe compartment when the lights dimmed, they began to behave more and more as if they had not been conditioned. The human subjects who could turn down the noise had an increased sense of control and could tolerate a louder sound before responding (i–j–k).

Critics of applied catastrophe theory sometimes contend that it does nothing which a simple 'threshold' model for the same phenomena cannot do. But as this model shows, it is misleading to think of a quantitative threshold in terms of only one variable, because the level of stimulus needed to evoke a given response may vary with some other factor. By setting forth the structurally stable ways in which this variation can happen, catastrophe theory again adds a new dimension to our view of behaviour patterns.

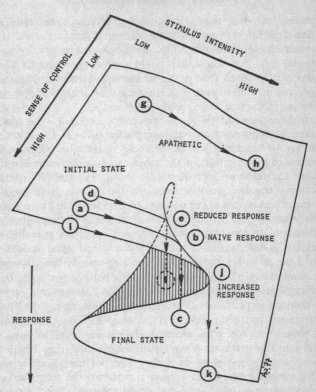

Figure 31. Learned helplessness and conditioned response

REACTIVE AND PROCESS SCHIZOPHRENIA

The victim of schizophrenia shows greatly impaired thought processes and often suffers terrifying delusions, although (unlike the psychotic patient) he usually retains his orientation of personal identity, place and time. There is strong evidence that the biochemistry of the brain plays a major role in schizophrenia. This evidence comes from recent studies of the neurotransmitters which convey nerve impulses across the tiny gap between nerve cells. Even before the details of neurotransmitter action began to emerge, an important clue had come from studies of identical twins : if one twin becomes schizophrenic, the other stands a 50 per cent chance of doing so, while in non-identical twins and in ordinary siblings that chance is only 10 per cent. It is generally agreed that most or all cases of schizophrenia involve a genetic predisposition to the disease, presumably through altered brain chemistry.

Two broad categories of schizophrenia are the *reactive* type, which appears in response to a particular stress or traumatic event in adult life, and the *process* type, in which mental abnormality begins early in life and becomes progressively more severe. In reactive schizophrenia, the onset of the disease comes as what one author calls a 'big bang', and a schizophrenic episode can end as suddenly as it begins. Process schizophrenics, on the other hand, begin to shun contact and social integration in late childhood, often develop delusions in adolescence, and are much more likely to be permanently disabled by the disease. There are also mixed or 'borderline' cases, in which the victim, when under stress, experiences mild confusion and thought impairment, but not necessarily clear-cut schizophrenic behaviour.

One way of explaining the difference between these types of schizophrenia is to distinguish particular stresses in adult life from what is called 'life-history stress', the cumulative psychological effect of many external influences. One such influence could be the conditions of development before birth : the same research which brought out the genetic factor in schizophrenia showed that of two identical twins, the smaller is more prone to develop the disease. Low birth weight might reflect foetal anoxia or abnormalities in the placenta which would affect brain development and neural chemistry. Another influence in life-history stress might be family strain during childhood, giving rise to the emotional traumas of Freudian analytic theory.

Figure 32 shows a model of reactive schizophrenia, with genetic predisposition and stress as the control factors. It indicates that a person without a strong genetic predisposition,

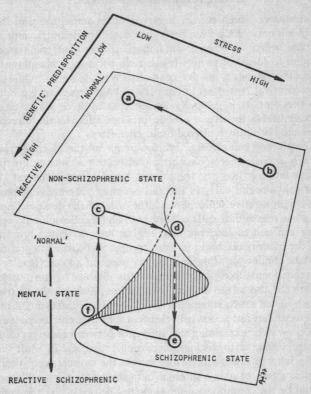

Figure 32. Reactive schizophrenia

while affected by stress to some extent, will not undergo the discontinuous shift to a schizophrenic mental state, and returns to normal when the stress is relieved (*a–b–a*). When there is a high genetic predisposition, though, the transitions provoked by stress and relaxation become drastic (*c–d–e* and *e–f–c*). Although quantitative assessments of stress are difficult, this model's suggestion of hysteresis behaviour is in accord with clinical experience for reactive schizophrenics whose occupation or home life is unremittingly stressful. They are caught in a vicious circle when the reduced stress of the hospital leads to a return to normal and they are released, only to encounter the same pressures again.

To model process schizophrenia, additional control factors – the sources of life-history stress – are needed. Figure 33 uses the butterfly catastrophe surface, employing the same control

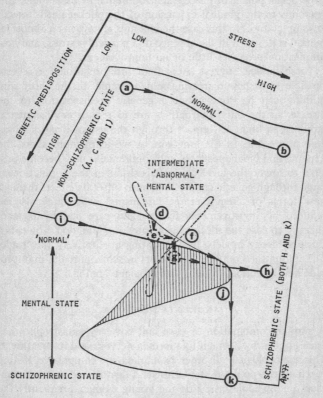

Figure 33. A more complex model using the butterfly catastrophe graph to illustrate both reactive and process schizophrenia

factors plus two more: pre-natal and childhood stresses. For some values of life-history stress, the behaviour surface is qualitatively identical to the cusp catastrophe surface: the normal response to stress (a–b) and that of the reactive schizophrenic (i–j–k) are possible. But at higher levels of life-history stress, the surface folds into the three-layer butterfly 'pocket', so that the intermediate, 'borderline' behaviour becomes possible. The slow development of process schizophrenia, in this model, corresponds to the gradual expansion of the pocket until it swallows up point c. In this case the victim is always in an abnormal mental state, even in the absence of a particular stress, and hospitalization produces little or no improvement.

These models suggest a coherent interpretation for several possible strategies in treating schizophrenia. In the short run, nothing can be done about a genetic predisposition to the disease, so the only alternatives are to reduce environmental stress or to blunt its impact through the use of drugs such as chlorpromazine. In the long run, as we learn more about the genetics and biochemistry of schizophrenia, it may become possible to lower the level of predisposition in the population at large through genetic counselling, or to offset its effects through new forms of chemotherapy. Life-history stress may also be reduced by improvements in pre-natal care and by psychotherapy to ease the tensions generated in childhood. The catastrophe-theory models, by illustrating qualitatively the 'trade-off' among the contributing factors in schizophrenia, may provide new directions in family and social planning.

ANOREXIA NERVOSA

A similar combination of cusp and butterfly catastrophes has been applied by Zeeman to anorexia nervosa and its treatment. This psychological disorder is marked by compulsive fasting (even to the point of starvation and death), and is most common among adolescent girls and young women. Frequently the anorexic patient begins with stringent dieting and progresses to a phase of alternate fasting and gorging, losing touch with the normal promptings of hunger and its satisfaction. This bimodal behaviour suggests the cusp catastrophe and hysteresis, which Zeeman modelled with eating as the behaviour axis and hunger and the progressive abnormality of appetite as the control factors. A British psychotherapist, J. Hevesi, collaborated with Zeeman in modelling Hevesi's unusually successful treatment for anorexia nervosa. The therapy involves repeated sessions of hypnotic trance during which the patient is reassured. The butterfly catastrophe model introduces two new factors: drowsiness which weakens the patient's obsessive concern with either resisting or giving in to appetite, and reassurance, which strengthens her self-confidence. Changes in these additional control factors bring about the creation of the intermediate pocket of the butterfly catastrophe surface, corresponding to the trance state, and its expansion until it opens on to the surface corresponding to normal (neither fasting nor gorging) behaviour. 'One of the strengths of the catastrophe-theory model of anorexia', Zeeman believes, 'is that it explains the patient's own description of herself. The seemingly incomprehensible terms in which some anorexics describe their illness turn out to be quite logical when viewed in the framework of the catastrophe surfaces.'

PSYCHOACOUSTICS

One of the most highly developed quantitative branches of psychology is psychophysics, which relates the measurable intensity of sensory stimuli to our ability to detect and to discriminate among them. Psychoacoustics, which combines the physics of sound waves with the psychology of hearing, reveals, for example, that the ability to tell which of two sounds is higher-pitched varies with frequency: we can make finer distinctions near the middle of the range of audible sounds than we can near the upper and lower limits. For some reason, while optical illusions have always played a large part in theories of vision, auditory illusions have only recently attracted comparable attention from students of psychoacoustics. An auditory illusion reported by Victor Hill, a mathematician who is also a concert harpsichordist, provides an example of discontinuity in perception which might be approached through catastrophe theory.

Hill has absolute pitch, the rare ability to recognize a note and to sing accurately a prescribed note. It extends to groups of notes, too : he can name the tones in a chord of six or ten notes sounded on the piano. He believes the ability to be an acute tonal memory which functions analogously to colour discrimination. Both absolute pitch and the conventions of musical composition and performance are relayed to physical sound frequency. As Hill points out, the standard reference frequency today is 440 Hz (cycles per second), equivalent to the A below middle C. In the early eighteenth century, though, A was equivalent to a standard frequency of about 415 Hz, one tone (or one half step) lower. Some harpsichords, like Hill's, allow the performer to shift the keyboard so that early compositions can be played as their composers intended.

Hill reports that when he plays his harpsichord set to the lower standard, he hears 415 Hz as A, but when he plays a normal piano, he hears 440 Hz as A. If someone else plays a composition Hill is familiar with on the harpsichord, Hill hears 415 Hz as A – but if the composition is unfamiliar, he hears the same frequency as A-flat, even though he knows that the keyboard is transposed. 'Even more curiously,' he says, 'it happened that a former student began to play a Scarlatti sonata in F-minor on my harpsichord when it was set at A-415 Hz. At first I heard it in E-minor, but the instant I recognized it, I had the sensation of a switch-over, and I began to hear it in F-minor.'

Figure 34 illustrates Hill's experience of discontinuous perception as a transition determined by his expectation. Like any other musician today, he has been trained on keyboards set to the standard frequency, but his knowledge of his own instrument alters what he *expects* to hear. The moments at which the transitions (b–c) and (d–a) occur depend upon whether or not he knows the composition. The curve in Figure 34 could be a

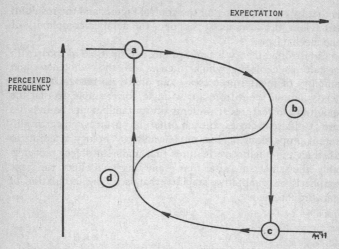

Figure 34. Perception and expectation: an auditory illusion

cross-section of the cusp surface, which would be expected if
this auditory illusion is controlled by only two factors – but
systematic experiments would be needed to verify that. The
curve might instead turn out to be part of a higher-dimensional
catastrophe surface, with a wider variety of smooth and discontinuous transitions, and in that case it might well lead to the
discovery of new auditory phenomena.

We have all had the experience of the solution to a nagging
problem jumping into the mind, or that of forgetting a name
that was on the tip of the tongue moments before. These sudden changes, too, might be modelled on a catastrophe surface
where the more customary, 'step-by-step' mental processes follow smooth paths. At the moment, although we are learning a
great deal about the neural and biochemical mechanisms of the
brain, we can only speculate on how they are co-ordinated in
large-scale mental activities such as perception and memory.
The advantage of catastrophe theory for descriptive purposes

is that by concentrating on the control factors and the resultant behaviour, it allows us to pass over the detailed complexity of the mechanisms.

Catastrophe theory is perhaps most compatible in spirit with gestalt psychology, which focuses on the wholeness and stability of mental processes – our ability to recognize an object when we see it from a new angle, for example, so that the quantitative features it presents are unfamiliar (its boundaries are at different angles to each other, its surfaces appear in different proportions, and so on). Gestalt psychology attempts to abstract the qualitative features that make an object recognizably 'the same' time after time – and this is, in effect, the same emphasis on qualitative stability that is at the foundation of catastrophe theory.

10

THE SHAPE OF
REVOLUTION

*'I do not know what I may appear to the world; but
to myself I seem to have been only like a boy play-
ing on the seashore, and diverting myself now and
then finding a smoother pebble or a prettier shell
than ordinary, whilst the great ocean of truth lay
all undiscovered before me'* – ISAAC NEWTON

MOST of us, most of the time, play it safe. We talk brightly of
the revolutionary changes around us, and yearn more or less
darkly for law and order. Like children in the surf, we enjoy
being tumbled because we know we can get back to the sand.
But a few of us, with enough insight and curiosity, play the
quiet games of 'How?' and 'Why?' Instead of splashing and
shouting, those few look for patterns in the sand, in the sea
and in the sky, and they sit and think. Yet when they are done,
it is as if a great wave had broken over us all and changed the
whole shore.

Newton might have been less modest had he foreseen the
changes his ideas would bring about. Such foresight, however,
is even rarer than genius itself. In *The Structure of Scientific
Revolutions*, the most influential modern work on the subject,
Thomas Kuhn introduced the concept of the scientific 'paradigm' – the implicit scheme within which theories are set forth
and research is conducted. We must remember, he cautions,
'how very limited in scope and application a paradigm can be
at the time of its first appearance'. Yet as time goes by, it determines the scientific questions that are asked, the way in which
answers are sought, and even the rules of evidence.

Catastrophe theory (in particular its essential concept of
structural stability) is really a paradigm rather than a theory. It

has attracted so much attention and generated so much argument because its scope and application appear to be virtually unlimited. But so much hangs on the phrase 'appear to be' that one should distinguish, as physicist Michael Berry does, among the different ways the theory has been applied to date. First, there are the true applications of the theory – those such as Berry's own work in optics, where the theory has made correct predictions and led to advances in understanding. Second, there are the illustrations – cases in which the theory produces in a new way results that have already been obtained by existing methods (as Thompson and Hunt realized when they correlated catastrophe theory with their analysis – via bifurcation theory – of elastic buckling). Third, there are what Berry calls 'invocations' of the theory – those in which the identification of the potential and the control factors remains tentative, so that the theory is employed 'because of the suggestiveness of its images in the hope that its axioms might eventually be shown to apply'. The third category takes in nearly all the uses of catastrophe theory to date in biology, and all of those in the social sciences and psychology.

For Thom himself, the scientific applications of his theory are secondary to its formal beauty and power, and its scientific status rests on its 'internal, mathematical consistency, which allows making deductions, generating new forms from another set of forms, thus allowing in favourable cases qualitative predictions, and in general realizing a considerable "reduction of arbitrariness" in the description'. By now, however, the theory has to a large extent escaped its creator. Although Thom considers it a language, an art of models, a theory of analogy, catastrophe theory is being tested, attacked and defended in terms appropriate to a scientific theory. And while most observers, even those who decry the theory's applications, agree that Thom's mathematics are both beautiful and consistent, they do not agree on the theory's scientific status. That judgement depends on the answers to a different set of questions:

Is catastrophe theory correct? In its mathematics, yes; in the natural philosophy that inspired it and the scientific applications that flow from it, the only possible answer is that it's too soon to say. There is always a chance of error whenever we try to capture any aspect of reality in mathematical symbols, and another chance of error when (after working with the symbols) we use them to generate descriptions or predictions of reality. But Thom's mathematics are very general; the assumptions required to apply them are few and reasonable; and the descriptions and predictions they make possible seem to agree with a great deal of what we already know. It is hard to believe, for example, that the group of behaviour types associated with the cusp – bimodality, divergence, continuous and discontinuous pathways of change, and hysteresis – are so often characteristic of the same natural processes purely by chance.

Is the theory useful? In the rigorous applications, yes; in the illustrations, sometimes; in the 'invocations', both yes and no. Yes, because catastrophe theory provides a common vocabulary for features of many different processes. Someday it may be as natural to speak of a 'cusp situation' or a 'butterfly compromise' as it is today to speak of the 'point of diminishing returns' or of a 'quantum jump'. No, because when the theory is invoked for the suggestiveness of its images, it cannot usually tell us anything we did not know before (although it can make explicit certain features that other models tend to neglect). As Thom says of qualitative modelling in social science, even if we obtained experimental confirmation, such confirmation 'would not be at all surprising, since the model is constructed precisely to generate the given morphology'. On the other hand, catastrophe theory will *become* useful in social science to the extent that it prompts social scientists to ask new questions – a process that is already beginning – and to the extent that its applications can be made rigorous, even quantitative – a process that will take some time.

Is the theory over-ambitious? Yes, of course it is – just like any other theory, whether scientific, social or philosophical. As Alfred North Whitehead put it, 'There are no whole truths; all truths are half-truths. It is trying to treat them as whole truths that plays the devil.' Catastrophe theory is still very new, and unquesionably its promise is still greater than its solid achievements. Newton's 'great ocean of truth' is still out there, all undiscovered; passing judgement on catastrophe theory's potential impact today is like trying to estimate the size of a wave that is still on the horizon. All we can say for sure is that this wave began very deep in the ocean, that so far it has travelled much faster than most, and that those are the attributes of what we call tidal waves, the ones that sooner or later affect us all.

Perhaps, as the critics imply, mathematicians like Thom and Zeeman should not be so confident that their ideas correspond to reality. Zeeman himself fancifully draws a cusp catastrophe surface (Figure 35) to illustrate the difference between sense and nonsense in mathematics. While most mathematicians follow the course $a–b$, Thom chose a different course. He began with a highly speculative idea (that of structural stability in nature), combined his own mathematics and that of others since Poincaré to contain it, and then attempted to express it with increasing rigour. Zeeman and many others believe that Thom made the 'jump' ($c–d–e$) years ago, and that the rest of us are only beginning to follow him.

However new Thom's ideas may seem, they are really part of a very old enterprise. They go back far beyond Poincaré's time; back beyond Newton's contemporary Leibniz, who said that 'as God calculates, so the world is made'; back even beyond Plato and Pythagoras, who believed that in the abstractions of number and geometry we are paradoxically closest to the nature of things; back to the first natural philosophers of whom we know anything, the Ionian Greeks of the fifth and sixth centuries B.C. 'The old idea of Man, the microcosm, mirroring World, the macrocosm, retains all its force,' Thom writes. For him, the question is not how our minds can conceive patterns applicable to the world, but how they could fail to do so, for we are in and of the world. Western mathematics, philosophy and science alike began with those questioners on

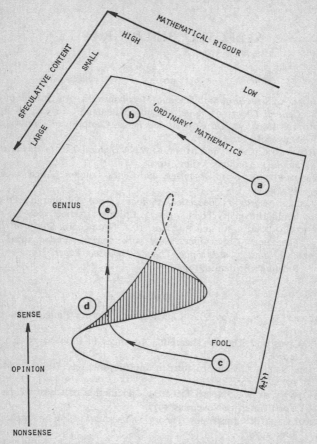

Figure 35. Sense and nonsense in mathematics. Where is catastrophe theory?

the shore of Asia Minor more than twenty-five hundred years ago. Perhaps more than anyone else alive today, René Thom is their rightful heir.

SELECTED BIBLIOGRAPHY

BOOKS

THOM, R., *Structural Stability and Morphogenesis: An Outline of a General Theory of Models*. Reading: Benjamin, 1975.

THOMPSON, D. W., *On Growth and Form*. Cambridge: Cambridge University Press, 1917.

THOMPSON, J. M. T., and HUNT, G. W., *A General Theory of Elastic Stability*. London: Wiley, 1973.

WADDINGTON, C. H., *Organisers and Genes*. London: Allen & Unwin, 1940.

(ed.), *Towards a Theoretical Biology*. 4 vol. Edinburgh: Edinburgh University Press, 1968–72; Chicago: Aldine, 1968–72.

WOODCOCK, A. E. R., and POSTON, T., *A Geometrical Study of the Elementary Catastrophes*. New York: Springer-Verlag, 1974.

ZEEMAN, E. C., *Catastrophe Theory, Selected Papers 1972–1977*. Reading: Benjamin, 1977.

ARTICLES

BERRY, M. V., 'Waves as Catastrophes'. *Physics Bulletin*, March 1976.

'Waves and Thom's Theorem'. *Advances in Physics*, 25, No. 1 (1976).

and NYE, J. F., 'Fine Structure in Caustic Junctions'. *Nature*, 5 May 1977.

BROWNE, M. W., 'Experts Debate the Prediction of Disasters'. *New York Times*, 19 November 1977.

CROLL, J., 'Is Catastrophe Theory Dangerous?' *New Scientist*, 17 June 1976.

DICKSON, D., 'Was Newton's Apple a Cusp or a Swallow-Tail?' *The Times Higher Education Supplement*, 5 December 1975.

DODSON, M. M., and HALLAM, A., 'Allopatric Speciation and the Fold Catastrophe'. *American Naturalist*, May/June 1977.

EKELAND, I., 'La Théorie des Catastrophes'. *La Récherche*, September 1977.

SELECTED BIBLIOGRAPHY

GOODWIN, B. C., 'Mathematical Metaphor in Development' (review). *Nature*, 16 March 1973.

GORMAN, J., 'The Shape of Change'. *The Sciences*, September/October 1976.

GUCKENHEIMER, J., Review of Thom's book, American Mathematical Society *Bulletin*, September 1973.

KAC, M., 'Some Mathematical Models in Science'. *Science*, 7 November 1969.

KILMISTER, C. W., 'The Concept of Catastrophe' (review). *The Times Higher Education Supplement*, 30 November 1973.

KOLATA, G. B., 'Catastrophe Theory: The Emperor Has No Clothes'. *Science*, 15 April 1977 (correspondence 17 June, 26 August).

KOZAK, J. J., and BENHAM, C. J., 'Denaturation: An Example of a Catastrophe'. *Journal of Theoretical Biology*, 63 (1976).

PAGE, J. K., JR, 'Death of a Theory'. *Smithsonian*, September 1977.

PANATI, C., 'Catastrophe Theory'. *Newsweek*, 19 January 1976.

ROSENHEAD, J., 'Prison "Catastrophe"'. *New Scientist*, 15 July 1976.

SCHULMAN, L. S., 'Tricritical Points and Type-Three Phase Transition:'. *Physical Review B*, vol. 7, no. 5 (1973).

STEEN, L. A., 'Catastrophe Theory: The First Decade'. *Science News*, 2 April 1977.

STEWART, I., 'The Seven Elementary Catastrophes'. *New Scientist*, 20 November 1975.

SUSSMANN, H. J., 'Catastrophe Theory'. *Synthèse*, 31 (1975).
'Catastrophe Theory: A Preliminary Critical Study'. *Proceedings of 1976 Biennial Meeting, Philosophy of Science Association*.
'Catastrophe Theory as Applied to the Social and Biological Sciences: A Critique'. *Synthèse*, 1978.

THOMPSON, J. M. T., 'Experiments in Catastrophe'. *Nature*, 3 April 1975.

WALGATE, R., 'René Thom Clears Up Catastrophes'. *New Scientist*, 4 December 1975.

ZAHLER, R. S., and SUSSMANN, H. J., 'Claims and Accomplishments of Applied Catastrophe Theory'. *Nature*, 27 October 1977 (correspondence 1 December, 29 December).

ZEEMAN, E. C., 'Catastrophe Theory. *Scientific American*, April 1976.

Many further references may be found in these sources.

INDEX

More About Penguins and Pelicans

SOCIAL SCIENCE LIBRARY

Manor Road Building
Manor Road
Oxford OX1 3UQ
Tel: (2)71093 (enquiries and renewals)
http://www.ssl.ox.ac.uk

This is a NORMAL LOAN item.

We will email you a reminder before this item is due.

Please see http://www.ssl.ox.ac.uk/lending.html
for details on:

- loan policies; these are also displayed on the notice boards and in our library guide.

- how to check when your books are due back.

- how to renew your books, including information on the maximum number of renewals.
  Items may be renewed if not reserved by another reader. Items must be renewed before the library closes on the due date.

- level of fines; fines are charged on overdue books.

Please note that this item may be recalled during Term.

Eurocommunism: Myth or Reality?
Edited by Paolo Filo della Torre, Edward Mortimer and Jonathan Story

Is Eurocommunism a sop to the West, and underneath, the same Stalinist wolf in sheep's clothing? Is it social democracy – a sheep in wolf's clothing? Or does it represent a genuine and momentous break from Moscow by the European Communist parties?

The analyses here, each written by an authority on his or her subject, cover virtually every conceivable aspect of Eurocommunism in an attempt to sort the reality from the myths. Part I focuses on the national parties in Italy, France, Spain and Portugal; their evolution and present predicament: Part II looks at the international implications of Eurocommunism, the problems the parties have in common and those they pose for the superpowers and the rest of the world.

Children in their Primary Schools
Henry Pluckrose

This book is written by an experienced headmaster for those who are interested in primary education in England but have little recent personal experience of primary schools. It examines both the content of the curriculum and the philosophy which underlies it, pointing out the different values of rigid curricular requirements in fixed age-grouped classes and family-group teaching with freer timetables – and showing us how play and work are equally essential, can be equally educative, and in many cases are actually interchangeable.

The NHS: Your Money or Your Life
Lesley Garner

Is the NHS terminally ill?

It is obvious from the increasing frequency with which the NHS hits the headlines that it needs major surgery. In her calm and comprehensive study, Lesley Garner shows that everywhere – both at home and abroad, regardless of economic or political bias – there is an ever more impossible struggle between supply and demand.

*Penguins for the best selection of Mathematics
and Statistics publications*

General Mathematics

Concepts of Modern Mathematics – Stewart
Geometry and the Liberal Arts – Pedoe
Mathematics and Logic – Kac and Ulam
Mathematics and the Imagination – Kasner and Newman
Mathematics in Management – Battersby
Mathematics in Western Culture – Kline
Newer Uses of Mathematics – Edited by Lighthill
Psychology of Learning Mathematics – Skemp
Riddles in Mathematics – Northrop
Mathematician's Delight – Sawyer

Recreational Mathematics

Further Mathematical Puzzles and Diversions – Gardner
Mathematical Puzzles and Diversions – Gardner
More Mathematical Puzzles and Diversions – Gardner
Mathematical Carnival – Gardner
Moscow Puzzles – Kordemsky

Statistics

Facts from Figures – Moroney
How to Lie with Statistics – Huff
How to Take a Chance – Huff
Lady Luck – Weaver
Statistics in Action – Sprent
Use and Abuse of Statistics – Reichmann

Calculators and Computers

Dictionary of Computers – Chandler
Electronic Computers – Hollingdale and Toothill

Forthcoming

Adventures with Your Pocket Calculator – Råde and Kaufman
Creative Use of Calculators – Killingbeck
Dictionary of Microprocessors – Chandor
Quick Statistics – Sprent
Statistics Without Tears – Rowntree
The Ambidextrous Universe – Gardner
Godel Escher Bach – Hofstader
The Penguin Book of Mathematical and Statistical Tables – Nelson
Reason by Numbers – Moore
What is the Name of this Book? – Smullyan